D1229597

DISCARD
T. W. PHILLIPS LIBRARY
BETHANY COLLEGE
BETHANY, WV 26032

DISCARD
LIBRARY

COAL SCIENCE

COAL SCIENCE

AN INTRODUCTION TO CHEMISTRY, TECHNOLOGY, AND UTILIZATION

Rita K. Hessley
John W. Reasoner
John T. Riley

Department of Chemistry
Western Kentucky University
Bowling Green, Kentucky

A WILEY-INTERSCIENCE PUBLICATION
JOHN WILEY & SONS
New York / Chichester / Brisbane / Toronto / Singapore

Library of Congress Cataloging in Publication Data:

Hessley, Rita K. (Rita Kathleen)
Coal science.

"A Wiley-Interscience publication."
Includes bibliographies and index.
1. Coal. I. Reasoner, John W. (John William), 1940– . II. Riley, John T. (John Thomas), 1942– . III. Title.

TP325.H43 1986 662.6'2 86-13
ISBN 0-471-81225-0

Printed in the United States of America

10 9 8 7 6 5 4 3 2 1

PREFACE

Coal is the most plentiful energy resource in the United States. It has been, and will continue to be, the principal fuel used by electrical utilities. As petroleum resources become more limited, the abundant coal resources will become more attractive for chemical feedstocks as well as sources of liquid and gaseous fuels.

Coal is a very complex and heterogeneous material. Efficient utilization of this resource requires an understanding of the formation, structure, and methods of analysis of the material. This text addresses each of these topics. It is an introductory-level text and requires a minimal background in chemistry and geology. The authors develop the topics as required. The goal of the text is to provide the reader with sufficient background to pursue independent study in coal science. Documentation and references are provided as additional sources of information for the reader and not as an exhaustive review of each topic.

Various parts of this text have been used in coal chemistry courses at Western Kentucky University and in professional short courses. Chapter 1 presents the chemistry and geology involved in the formation of coal. Chapter 2 presents the petrology and petrographic characterization of coal. The organic composition and structure of coal is covered in Chapter 3, while coal reactivity and the chemistry of coal conversion and utilization are the major topics discussed in Chapter 4. Chapter 5 presents a complete discussion of the routine testing and analytical characterization of coal. Overall the text should expose the reader to the more important topics of coal science so that a better understanding of the nature and utilization of our most abundant resource can be obtained.

RITA K. HESSLEY
JOHN W. REASONER
JOHN T. RILEY

Bowling Green, Kentucky
May 1986

CONTENTS

COAL SCIENCE

1
AN INTRODUCTION TO COAL AND COAL SCIENCE

"Coal takes first place among all useful minerals and rocks in world economics" (1). Today that statement may be considered by many scientists and economists to be brash, if not clearly unfounded. But few would deny that any other mineral, or rock, has had as much impact on world economics as coal.

Because of the recent renewed interest in utilizing the vast wealth of coal found lavishly distributed across the entire United States, this chapter is designed to establish the foundation for a student serious about understanding both the complexities and the potentials that coal holds for the future as a significant, if not the most significant, energy resource this nation has. It is essentially a survey of the historical record that clearly shows the influence of geology, chemistry, economics, politics, and even individual personalities on our understanding of what coal is.

The chapter is divided into 10 sections and presents significant definitions, a description of the nature of coal as a resource and the current status of our national reserves, the fluctuating character of coal production and coal consumption in the United States, previous and contemporary uses of coal, and a discussion of the organizations and official systems used to characterize and classify coal.

1.1 A SURVEY OF COAL THROUGH HISTORY

First among the reasons why coal may in fact be the most useful mineral, second only to food and water as a necessity of life, is that coal is one of the most ancient useful rocks. From earliest recorded history, material believed to be coal was in use in China. The Greek philosopher and teacher Aristotle mentioned coal in his writing 2,400 years ago (2). Coal has been referred to in English history for centuries, although it was not always referred to positively. King Edward I banned it as a domestic fuel in 1306. The interests of nobles—

the first environmentalists—prevailed in medieval England, and digging for coal by freemen was forbidden in the thirteenth century in order to protect game lands. The introduction of brick chimneys once again permitted its general use 100 years later, only to lead to the imposition of burdensome coal-use taxes ("hearth money") in the mid-sixteenth century in England and in France (3).

Coal "research" was also under way centuries ago. Aristotle recognized coal as having fossil components, and in 1546 Agricola promulgated his erroneous theory that coal came from oil. Several other theories of coal formation have been developed over the years, as we shall see later (4).

Coal exploration and trade thrived in England at the time North America was colonized. Native American Indians certainly used coal much earlier in the midwest and west, as well as in the eastern North American continent. Coal was discovered by colonists in Canada in 1672 and in the American colonies in 1673. The first mine was opened in the area that is now Pittsburgh in 1684. Seams in eastern Pennsylvania were not uncovered for nearly another century (5). Documents show that the newfound freedom of private ownership flourished even before independence was won when "all the known bituminous coal fields in Pennsylvania" were bought from six Indian nations by the Penn Proprieties Company (6). The invention and development of the steam engine in the eighteenth century had a spectacular impact on the use of coal. In fact, the entire industrial revolution relied almost exclusively on power derived from coal.

In 1918, after 130 years of large-scale productive commercial mining, the United States provided 579 million short tons of coal to industrial and domestic consumers (one metric ton consists of 2,200 lb, and 1 short ton is equivalent to 2,000 lb). That level could not be maintained during the Depression, and in 1937 the production level dropped to 495 million short tons. World War II spawned a renewed demand that pushed production to a record level, at that time, of 630 million tons by the middle of the decade (7). Since 1945, United States coal production has passed through two 10-year phases. In the decade from 1951 to 1961, there was a steady decline in production to 427 million short tons. This downward trend was reversed during the next decade, when output rose to near prewar levels, with 565 million short tons produced in 1970 (8). In both cases, the ebb and flow of competition among the various sources of available energy explain the importance of coal in the recent past.

By 1970, nuclear power had become a practical reality. At about the same time, increased limitations were imposed on the amount of undesirable emissions allowed from coal-fired power facilities, and the future of the coal industry in the United States became very bleak. However, the dramatic expansion of strip mining operations (from 1.5% of all coal mined in 1920 to 47% in 1973), more economic coal transportation, and diminished reliance on imported fuel supplies all helped to maintain coal production and utilization at a competitive level.

Since 1970, then, coal production has fluctuated rather than showing a clear trend either up or down. The United States—as well as most other countries that have large-scale technological demands for energy and whose level of energy consumption increases geometrically—recognized that a serious and extensive analysis and reevaluation of the energy potential of our coal resources was imperative regardless of the potential of nuclear sources.

This effort has not been a trivial exercise in measuring known mine acreage and projecting industrial and population growth and demand patterns. In fact, the undertaking has involved the highest levels of national and international government, a multitude of state and local agencies, corporate and private research facilities, millions of dollars nationwide, and probably billions of dollars worldwide. Some aspects of the survey and analysis have been completed. Population growth patterns and energy consumption levels have been characterized and projected. Operational mine capacity and the quality of the coal produced has been evaluated in the light of current and expected environmental and engineering process demands. The search for additional coal resources and the development of safer and more economical ways to use coal are only now beginning to take shape. We are coming full circle. The high level of technology that spawned the need for intensive cultivation and development of alternatives to coal is now helping the nation address and solve the problems associated with rediscovering coal as one valuable way to satisfy our current and long-range needs.

1.2 DEFINITIONS OF COAL

Before we begin a discussion of the coal resources in the United States, it is important to establish what sort of material has been defined as coal. Although the ordinary layperson, or even a schoolchild, would probably give a reasonable physical description of coal, to scientists in different disciplines, giving a suitable description of coal is not necessarily a simple task and can cause debate among geologists, chemists, and engineers. G. L. Jansen, a geologist and coal petrographer, refers to coal quite simply as an "organic sediment" (9). A similar description is that "coal is a very complex heterogeneous mixture of organic compounds and minerals . . . [an] analogy . . . to a fruit cake" (10). This kind of description is frequently used by geologists to distinguish coal from hard rock minerals. For contracts, licenses, and other legal purposes, coal is considered a mineral (11). Strictly defined, a mineral is inorganic, homogeneous, and of definite composition. Scientists dealing primarily with coal and its end uses prefer a more explicit definition. More specifically, it can be said that coal is a combustible solid, usually stratified, which originated from the accumulation, burial, and compaction of partially decomposed vegetation in previous geologic ages. Because the layers of vegetation have undergone varying degrees of chemical and physical alteration, or metamorphosis, coal exhibits a wide range of composition and chemical and physical properties (12). In this sense, oil shale and other carbonaceous shales are excluded because they contain a proportionally larger amount of inorganic material. Likewise, peat, although it is the immediate predecessor of coal, is considered separate from coal because it has not undergone metamorphosis. There is yet another group of deposited organic solids that can be distinguished from coal. These materials are called bitumens. They are derived from petroleum and do not contain discernable plant remains. The term *bitumen* has unfortunately also been used to describe organic compounds that can be extracted from peat and coal with solvents such as benzene. Furthermore, the primary products of destructive distillation of coal are also called bitumens. This material is a complex mixture of resins, waxes, and a variety of hydrocarbons (13). A discussion of all these materials and the techniques of solvent extraction is presented in Chapters 3 and 4.

Chemically, coal consists primarily of carbon, hydrogen, and oxygen, with lesser amounts of as many as 60 other elements. Some of the elements in coal come from the original plant matter, while others have been incorporated into the coal's framework during the coalification process or have accumulated as extraneous matter in the mining and preparation processes.

1.3 COAL RESOURCES IN THE UNITED STATES

The location of the country's coal fields and the abundance and availability of the coal they contain was a critical factor in past history in determining the way population and industrial centers were distributed across the nation. Today the discovery of new coal resources is not as critical as the nature of the coal and whether we have the means to use it in an economically and environmentally sound manner.

1.3.1 Geological Character of Coal-Bearing Rocks

Although Lindberg and Provorse state that some coal has been found in almost every state in the union (6), it is generally accepted that not every state has deposits of commercially valuable coal. Mining operations are carried on in only 26 states. Table 1.1 lists the states with coal-bearing areas and shows the land area of each that is underlain with coal. Thirty-seven states are included, and the total amounts to about 13% of our total land area. Despite the fact that coal deposition requires that a number of independent variables coincide in time, coal actually occurs in predictable locations across the country. Areas more amenable to coal deposition are those where sedimentary rock formations are found. These rocks were formed near the earth's surface from materials that collected and settled without geologic violence. Although we have defined coal as a material that has undergone degrees of chemical and physical alteration, it should be borne in mind that most organic material would not withstand extreme conditions of temperature or pressure without decomposing. Fossil fuels, then, including petroleum and shale, can be expected to be found under the vast expanse of land where the geologic formations are composed largely of sedimentary rocks. The particular type of material deposited, the climate at the

TABLE 1.1 Size and Percentage Distribution of Coal-Bearing Areas in the United States

State	Total Area of State (mi²)[a]	Area Underlain by Coal-Bearing Rocks	
		Square Miles	Percent
Alabama	51,609	9,700	19
Alaska	586,412	35,000	6
Arizona	113,909	3,040	3
Arkansas	53,104	1,700	3
California	158,693	230	.1
Colorado	104,247	29,600	28
Georgia	58,876	170	.2
Idaho	83,557	500	.6
Illinois	56,400	37,700	67
Indiana	36,291	6,500	18
Iowa	56,290	20,000	36
Kansas	82,264	18,800	23
Kentucky	40,395	14,600	36
Louisiana	48,523	1,360	3
Maryland	10,577	440	4
Michigan	58,216	11,600	20
Mississippi	47,716	1,000	2
Missouri	69,686	24,700	35
Montana	147,138	51,300	35
Nebraska	77,227	300	.4
Nevada	110,540	50	—
New Mexico	121,666	14,650	12
New York	49,576	10	—
North Carolina	52,586	155	.3
North Dakota	70,665	32,000	45
Ohio	41,222	10,000	24
Oklahoma	68,919	14,550	21
Oregon	96,981	600	.6
Pennsylvania	45,333	15,000	33
South Dakota	77,047	7,700	10
Tennessee	42,244	4,600	11
Texas	267,338	16,100	6
Utah	84,916	15,000	18
Virginia	40,817	1,940	5
Washington	68,192	1,150	2
West Virginia	24,181	16,800	69
Wyoming	97,914	40,055	41
Other states	312,855	0	0
Total	3,615,122	458,600	13

Source: Reference 14.

[a] U.S. Bureau of the Census, 1973, *Statistical Abstract of the United States*, 94th ed., p. 172.

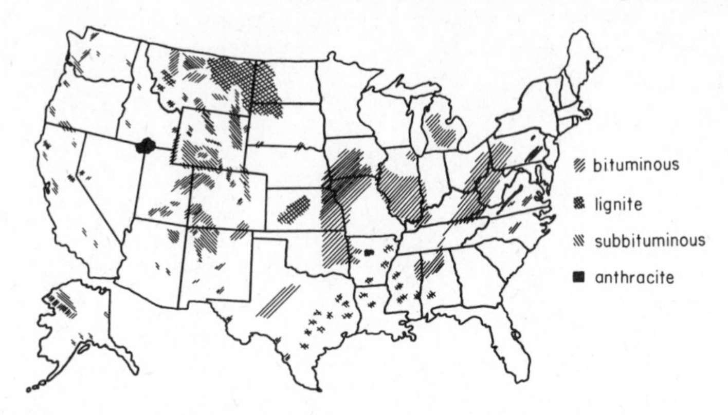

FIGURE 1.1 Major coal-bearing regions in the United States. (*Source:* Reference 16; reprinted with permission from *The Power of Coal,* National Coal Association, Washington, 1978, p. 13.)

time of deposition, and the chemical processes by which some of this sedimentary material was formed specifically into coal are presented in Chapter 2.

Coal may be found in layers alternating with stone or clay, or in isolated, single seams. The latter formations tend to be thicker but less extensive than the alternating layers. It has been reported that of all coal-bearing rock formations, coal constitutes only about 4% of the thickness. In the Pennsylvania–West Virginia area alone, however, there are 117 named coal seams (15). Figure 1.1 shows the regions in the United States where coal is found. These regions represent about one-fifth of the world's accessible coal resources. Including Canada, this figure rises to one third of the world's reserves.

1.3.2 Seam Character

The area, thickness, and depth of coal deposits all vary widely, and in some cases the extent has only been estimated within large limits. The Appalachian Highlands coal field near Pittsburgh has been described by various estimates as underlying an area between 10,000 and 30,000 mi^2 (17). Of coal seams around the world, three deposits in Victoria, Australia, have been found to be 166, 227, and 266 ft thick, and the main seam at Fusin, Manchuria, China, is the thickest known seam, measuring 400 ft thick in places (15). In the United States, seams range from a few inches to 100 ft. Most producing seams are between 2 and 10 ft thick, and the average thickness is 5.5 ft (18). In central Michigan, a relatively large deposit of coal is not mined because it is only a few inches thick throughout.

The depth at which a coal seam lies has important economic implications. Unless deeply buried coal is of extremely high quality, lies in an otherwise fuel-poor region, or is influenced by other economic factors, the cost of removing the overburden or of drilling and erecting shaft mine operations is prohibitively expensive. The deepest mine in the world is the Produits Colliery near

Mons, Belgium. It is mined at a depth of 3,900 ft. Some mines in England, Germany, and Australia are also more than 3,000 ft deep. For comparison, gold is mined in Brazil below 7,000 ft, and diamonds are mined in South Africa at depths up to 4,000 ft (19). In the United States, a seam is mined in eastern Pennsylvania at a depth greater than 2,200 ft, and the entire seam is estimated to extend as much as 2,300 ft farther downward. Most mining operations are carried out at much shallower levels, and some very deep shafts have been abandoned for economic reasons. In the western part of the country, some seams are buried under extensive overburden caused by folding and faulting of the Rocky Mountains, while others are much nearer the surface, permitting easy access for strip mining. Finally, at the extreme of minimum depth, coal crops out frequently at the surface, which probably accounts for its initial discovery as a fuel by prehistoric human beings. Outcrops can be part of a very thin seam near the surface or an upper extension of a seam thousands of feet deep (20).

1.4 COAL RESERVES AND RESOURCE ASSESSMENT

Several factors must be borne in mind when evaluating how much coal remains underground and in predicting how long that coal will last. First, exploration is not absolutely completed. New deposits may be discovered. Second, changing population growth patterns, economic factors, and rapidly advancing technological developments all add dimensions of uncertainty to any assessment of our available reserves of coal. Third, what is actually meant by *available* is also subject to change.

1.4.1 United States Geological Survey Assessment

The United States Geological Survey uses the following terms with regard to our coal resources (21):

Identified resources: coal-bearing rock whose location and existence is known

Resources: total quantity of coal in the ground within specified limits of bed thickness and overburden thickness; comprises identified and hypothetical resources

Hypothetical resources: estimated tonnage of coal in the ground in unmapped and unexplored parts of known coal basins to an overburden of 6,000 ft; determined by extrapolation from nearest area of identified resources; includes coal believed to be present in the continental shelves

Measured resources: tonnage of coal in the ground based on assured coal bed correlations and on closely spaced observations about one-half mile apart; computed tonnage judged to be accurate within 20% of true tonnage

Indicated resources: tonnage of coal in the ground based partly on specific observations, partly on reasonable geologic observations, and partly on reasonable geologic projections; points of observation and measurement about 1 mi from beds of known continuity

Inferred resources: tonnage of coal in the ground based on assumed continuity of coal beds adjacent to areas containing measured and indicated resources

Demonstrated reserve base: selected portion of coal in the ground in the measured and indicated category; restricted primarily to coal in thick and intermediate beds less than 1,000 ft below the surface and deemed economically and legally available for mining at time of determination

Recoverability factor: percentage of coal in the reserve base that can be recovered by established mining practices

Reserve: tonnage that can be recovered from the reserve base by application of the recoverability factor

Table 1.2 gives the latest available figures for the total estimated remaining resources in the United States. Care must be taken when interpreting such figures, and especially in trying to make comparisons with other data. The value of 3.9 trillion short tons of reserves shown in the table represents identified and hypothetical reserves up to a 6,000-ft overburden. A more realistic figure is the 483 billion short tons of demonstrated reserve base (DRB) of coal reported in 1984 (24). It was estimated that this quantity of coal would be sufficient for more than 425 years at the 1974 rate of consumption (23).

Table 1.3 shows how the demonstrated reserve base is distributed across the United States. Based on these figures, about 55% of the nation's DRB lies in the west, but only one-third of the DRB is accessible by strip mining (23). It has been suggested by a number of environmentalist groups that extensive strip mining operations in the Great Plains and Rocky Mountain provinces be carefully controlled if not curtailed. In order to produce as much heat from low-sulfur coal as is available from eastern bituminous coal, more low-heat coal must be consumed. The total sulfur output, then, may equal or even exceed that of the high-sulfur eastern coal (25).

1.4.2 Evaluation of Coal Resources

Coal deposits were formed over the course of thousands of years through the alternating processes of land elevation, plant growth and decay, and land subsidence. Because of this, coal is a highly variable and heterogeneous material. In one coal deposit in Nova Scotia, for instance, 68 levels of different plant types have been found across a 1,400-ft seam (26). It is true that all coal has essentially a common origin, and basically the same components in varying proportions, but it is not true that "coal is coal and nature has created no particular kinds" (27). Substantial and significant differences in coal samples can be mea-

TABLE 1.2 Total Estimated Remaining Coal Resources of the United States, January 1, 1974

State	Overburden 0–3,000 Ft							Overburden 3,000–6,000 Ft	Overburden 0–6,000 Ft
	Remaining Identified Resources, January 1, 1974					Estimated Hypothetical Resources in Unmapped and Unexplored Areas [a]	Estimated Total Identified and Hypothetical Resources Remaining in the Ground	Estimated Additional Hypothetical Resources in Deeper Structural Basins [a]	Estimated Total Identified and Hypothetical Resources Remaining in the Ground
	Bituminous Coal	Subbituminous Coal	Lignite	Anthracite and Semianthracite	Total				
Alabama	13,262	0	2,000	0	15,262	20,000	35,262	6,000	41,262
Alaska	19,413	110,666	[b]	[c]	130,079	130,000	260,079	5,000	265,079
Arizona	21,234[d]	[d]	0	0	21,234	0	21,234	0	21,234
Arkansas	1,638	0	350	428	2,416	4,000[e]	6,416	0	6,416
Colorado	109,117	19,773	20	78	128,948	161,272	290,220	143,991	434,211
Georgia	24	0	0	0	24	60	84	0	84
Illinois	146,001	0	0	0	146,001	100,000	246,001	0	246,001
Indiana	32,868	0	0	0	32,868	22,000	54,868	0	54,868
Iowa	6,505	0	0	0	6,505	14,000	20,505	0	20,505
Kansas	18,668	0	[f]	0	18,668	4,000	22,668	0	22,668
Kentucky									
Eastern	28,226	0	0	0	28,226	24,000	52,226	0	52,226
Western	36,120	0	0	0	36,120	28,000	64,120	0	64,120
Maryland	1,152	0	0	0	1,152	400	1,552	0	1,552
Michigan	205	0	0	0	205	500	705	0	705

TABLE 1.2 Total Estimated Remaining Coal Resources of the United States, January 1, 1974, (*continued*)

	Overburden 0–3,000 Ft							Overburden 3,000–6,000 Ft	Overburden 0–6,000 Ft
	Remaining Identified Resources, January 1, 1974								
State	Bituminous Coal	Subbituminous Coal	Lignite	Anthracite and Semianthracite	Total	Estimated Hypothetical Resources in Unmapped and Unexplored Areas [a]	Estimated Total Identified and Hypothetical Resources Remaining in the Ground	Estimated Additional Hypothetical Resources in Deeper Structural Basins [a]	Estimated Total Identified and Hypothetical Resources Remaining in the Ground
Missouri	31,184	0	0	0	31,184	17,489	48,673	0	48,673
Montana	2,299	176,819	112,521	0	291,639	180,000	471,639	0	471,639
New Mexico	10,748	50,639	0	4	61,391	65,556 [g]	126,947	74,000	200,947
North Carolina	110	0	0	0	110	20	130	5	135
North Dakota	0	0	350,602	0	350,602	180,000	530,602	0	530,602
Ohio	41,166	0	0	0	41,166	6,152	47,318	0	47,318
Oklahoma	7,117	0	[f]	0	7,117	15,000	22,117	5,000 [h]	27,117
Oregon	50	284	0	0	334	100	434	0	434
Pennsylvania	63,940	0	0	18,812	82,752	4,000 [i]	86,752	3,600 [j]	90,352
South Dakota	0	0	2,185	0	2,185	1,000	3,185	0	3,185
Tennessee	2,530	0	0	0	2,530	2,000	4,530	0	4,530
Texas	6,048	0	10,293	0	16,341	112,100 [k]	128,441	[k]	128,441
Utah	23,186 [l]	173	0	0	23,359	22,000 [m]	45,359	35,000	80,359
Virginia	9,216	0	0	335	9,551	5,000	14,551	100	14,651
Washington	1,867	4,180	117	5	6,169	30,000	36,169	15,000	51,169

West Virginia	100,150	0	0	0	100,150	0	100,150	0	100,150
Wyoming	12,703	123,240	[b]	0	135,943	700,000	835,943	100,000	935,943
Other states[n]	610	32[o]	46[p]	0	668	1,000	1,688	0	1,688
Total	747,357	485,766	478,134	19,662	1,730,919	1,849,649	3,580,568	387,696	3,968,264

Source: Reference 22.

Note: In millions (10^6) of short tons. Estimates include beds of bituminous coal and anthracite generally 14 in. or more thick, and beds of subbituminous coal and lignite generally 2½ ft or more thick, to overburden depths of 3,000 and 6,000 ft. Figures are for resources in the ground.

[a] Source of estimates: Alabama, W.C. Culbertson; Arkansas, B.R. Haley; Colorado, Holt (1975); Illinois M.E. Hopkins and J.A. Simon; Indiana, C.E. Wier; Iowa, E.R. Landis; Kentucky, K.J. Englund; Missouri, Robertson (1971, 1973); Montana, R.E. Matson; New Mexico, Fassett and Hinds (1971); North Dakota, R.A. Brant; Ohio, H.R. Collins and D.O. Johnson from data in Struble and others (1971); Oklahoma, S.A. Friedman; Oregon, R.S. Mason; Pennsylvania anthracite, Arndt and others (1968); Pennsylvania bituminous coal, W.E. Edmunds; Tennessee, E.T. Luther; Texas lignite, Kaiser (1974); Virginia, K.J. Englund; Utah, H.H. Doelling; Washington, H.M. Beikman; Wyoming, N.M. Denson, G.B. Glass, W.R. Keefer, and E.M. Schell; remaining states, by the author.

[b] Small resources of lignite included under subbituminous coal.

[c] Small resources of anthracite in the Bering River field believed to be too badly crushed and faulted to be economically recoverable (Barnes, 1951).

[d] All tonnage is in the Black Mesa field. Some coal in the Dakota Formation is near the rank boundary between bituminous and subbituminous coal. Does not include small resources of thin and impure coal in the Deer Creek and Pinedale fields.

[e] Lignite.

[f] Small resources of lignite in western Kansas and western Oklahoma in beds generally less than 30 in. thick.

[g] After Fassett and Hinds (1971), who reported 85,222 million tons "inferred by zone" to an overburden depth of 3,000 ft in the Fruitland Formation of the San Juan Basin. Their figure has been reduced by 19,666 million tons as reported by Read and others (1950) for coal in all categories also to an overburden depth of 3,000 ft. in the Fruitland Formation of the San Juan basin. The figure of Read and others was based on measured surface sections and is included in the identified tonnage recorded in table 2.

[h] Includes 100 million tons inferred below 3,000 ft.

[i] Bituminous coal.

[j] Anthracite.

[k] Lignite, overburden 200–5,000 ft; identified and hypothetical resources undifferentiated. All beds assumed to be 2 ft thick, although many are thicker.

[l] Excludes coal in beds less than 4 ft thick.

[m] Includes coal in beds 14 in or more thick, of which 15,000 million tons is in beds 4 ft or more thick.

[n] California, Idaho, Nebraska, and Nevada.

[o] California and Idaho.

[p] California, Idaho, Louisiana, and Mississippi.

TABLE 1.3 Distribution of Demonstrated Reserve Base of Coal across the United States

DRB	East [a]	West	Percent of Total
Anthracite	7	—	2
Bituminous	220	36	51
Subbituminous	0	182	37
Lignite	1	43	9

State [b]	Low-Sulfur DRB	Total DRB	Percent of Low-Sulfur Total
Pennsylvania	426	30,381	1
West Virginia	12,806	39,776	32
Virginia	2,254	3,471	65
Kentucky	5,321	34,002	16
Wyoming	36,147	69,924	52
Colorado	8,864	17,281	51
Montana	118,315	120,428	98
Utah	3,193	6,478	49
New Mexico	1,822	4,521	40

Source: Reference 23.

[a] In millions of short tons.

[b] Includes states with more than 2 billion short tons total DRB; values given are in millions of short tons.

sured. Ultimately, coal is not all the same. From the earliest industrial times it has been important to classify coal in descriptive categories. Unfortunately, a number of different systems have evolved together, and they are not unequivocal. However, some generally recognized and accepted terms are in use throughout the coal industry. It will be helpful to have some familiarity with the way coal is differentiated and some reference to the use of these differences in assessing the quality and usefulness of coal reserves.

Type, rank, and *grade* are three terms that describe particular characteristics of coal. The kinds of plant material from which a coal originated, the kinds of mineral inclusions, and the nature of biochemical reaction conditions that prevailed during periods of plant decay give rise to coal types. The two main types of coal are banded and nonbanded. As the name indicates, banded coal consists of discrete layers or bands composed in part of the remains of woody plant tissues, fine plant debris, and "mineral charcoal" or fusain. Smoother and more finely grained coals derived primarily from spores and algae constitute nonbanded coal (28),

The rank to which a coal is assigned refers to the extent of metamorphosis the deposited plant and animal matter has undergone. The pressure and heat caused by protracted periods of burial or by the folding of the earth's crust brought about progressive changes in the structure of the organic material as it was converted into coal. Coal that has undergone the most extensive change, or metamorphosis, has the highest rank. Chemical analyses are carried out to determine the degree of metamorphosis in a sample. In practice, the rank of a coal is assigned according to its fixed carbon value or its heating value. The fixed carbon value is obtained by subtracting the moisture, ash and volatile matter from 100. Ash is the residue that remains after combustion of a coal sample, and volatile matter is the material driven off by heating a coal sample in the absence of air in a special furnace under prescribed conditions. Volatile matter includes all the gases, other than water vapor, that are evolved under the conditions of the test. It has been found that the process of coalification increases the amount of fixed carbon and generally decreases the amount of moisture and the amount of volatile matter found in coal. The coalification series begins with lignite as the lowest-rank coal and progresses with gradual changes in fixed carbon and volatile matter to anthracite, or hard coal, which has the highest rank. These gradual changes are illustrated in Figure 1.2.

Although all coal, even at the highest rank, contains moisture, the amount of moisture in a sample of coal is extremely important to its value to virtually

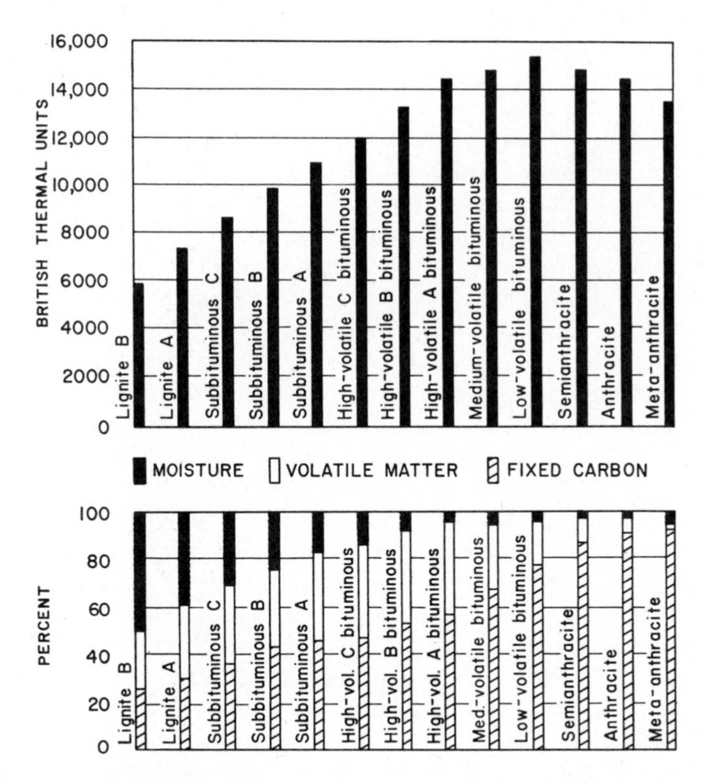

FIGURE 1.2 Variations in heat content, fixed carbon, volatile matter, and moisture with rank of coal. (*Source:* Reference 29.)

every user. First, for every unit weight of water in a shipment of coal, less coal is being received. Since the price is paid per total unit (usually per ton) of material, the presence of more than a small percentage of moisture in coal from a particular area renders it of less commercial value. Second, once coal is ignited, the amount of heat it gives off also determines its relative value for uses ranging from domestic heating fuel to boiler fuel in steel or power production. The conversion of water from a liquid to a vapor requires a substantial amount of heat—heat that is "stolen" from the coal itself, again diminishing its value as a fuel. The heat content, calorific value, or energy value of coal can be determined in a laboratory test and is recorded in British thermal units (Btu). One Btu is the amount of heat required to raise the temperature of 1 lb of water 1° F. As the rank of coal increases, so does the ratio of fixed carbon to moisture, and the potential heat content, or energy value, of the coal. Figure 1.2 also shows the relationshiip between the rank of a coal and its energy content. Notice that the highest-rank anthracite coals have somewhat less heating value than do the medium- and low-volatile bituminous coals. This illustrates the complex chemical nature of coal and shows that, while the amount of fixed carbon may be used to establish the rank of a coal, the amount of volatile matter and moisture cannot be overlooked. A complete description of these terms and the manner in which they are determined is given in Chapter 5.

The grade of a coal refers to the amount and the kind of inorganic mineral impurities found bound into the coal. Among a variety of inorganic components, sulfur is perhaps the most significant. It interferes with catalytic materials used in coal conversion processes and is converted to gaseous sulfur dioxide, SO_2, a toxic and corrosive pollutant, when coal is burned. Although the kinds and amounts of inorganic components are becoming increasingly critical as ecological and environmental considerations, the rank to which a coal belongs is more frequently referred to than its grade in general discussions.

Before leaving this topic of the character and value of coal, mention should be made of the coal identified as "meta-anthracite" in Figure 1.2. Meta-anthracite, sometimes called superanthracite in older texts, is organic material that has metamorphosized almost completely to graphite. It is not found in many locations and, because it has a diminished heat content, is not a very valuable coal. The Narragansett Basin region of Rhode Island and Massachusetts contains an extensive field of meta-anthracite, probably the only such deposit in the United States. It is estimated that there are 400 million tons of meta-anthracite coal in the 900-mi^2 basin, and mining was carried on there until about 1920. Ironically, perhaps, the region is now the focus of renewed geologic investigation of its current commerical value as a source of graphite as well as for its coal. The effects of previous mining operations throughout the basin are also of growing concern because a number of residential and commercial areas have been damaged recently when land caved in over unmapped abandoned mine sites. Extreme sheer forces—folding and thrusting—associated with the collision of North America with West Africa or with South America are responsible for the extreme degree of metamorphosis.

1.5 COAL PROVINCES IN THE UNITED STATES

An examination of Table 1.1 and Figure 1.1 shows that coal is spread across the entire United States and that its location corresponds closely to the areas of sedimentary rock. Illinois and West Virginia are unique in that their coal fields are so pervasive that two-thirds of the total land area of each state is underlain with coal. In general, eastern coal is older and of higher rank than western coal. Except for the special case of the Narragansett Basin in Rhode Island and Massachusetts, coal is not found along either of our coasts. For reference purposes, the United States Geological Survey has divided the nation into six geographical coal provinces (30).

The Eastern Province extends from the Rhode Island coast south-southwest into Alabama. Recent figures that exclude the Narragansett Basin show that this area contains about 14% of the nation's coal and produced slightly more than 48% of the nation's coal in 1983 (31). Virtually all of the anthracite mined worldwide is found in eastern Pennsylvania, although small anthracite deposits are known in several other states (32). Except for small amounts of lignite located in southern Alabama, the remainder of this province contains only high-quality bituminous coal. Unexcelled as the source of domestic and industrial fuel, the Eastern Province, with only 11% of the country's gross coal reserves, has yielded more than 78% of all the coal mined in the United States.

The Interior Province includes a portion of Michigan (which is no longer mined), Illinois, Indiana, western Kentucky, Iowa, Missouri, Kansas, Oklahoma, Arkansas, and north-central Texas. This region produced 23% of the 1983 total tonnage. Generally, the coal found throughout this region is bituminous rank. Coal from much of this region also contains a high volatile content and sufficient sulfur to render it much less suitable than eastern and western coals as boiler fuel or for coke production.

The extensive lignite deposits that stretch northeastward from central Texas through central Arkansas, Louisiana, and Mississippi and into southern Alabama constitute the Gulf Province. Although, in 1929, W.T. Thom, a geologist then at Princeton University, reported that large-scale mining operations were under way in Texas at that time, more recent production statistics do not include Texas as a producer of lignite until after 1964 (33). Lignite is the lowest-ranking coal. Its high moisture content (35–45%) and correspondingly low heat content (6,300–8,300 Btu/lb versus 15,000 Btu/lb for high-quality bituminous coal), plus its tendency toward rapid crumbling (slacking), decrepitation, and spontaneous combustion, make it less desirable for many uses. Interest in lignite is growing, however, because coal conversion technology, especially for coal gasification, is progressing, and particular attention is being paid to the utilization of lower-quality coals.

The Great Plains states, consisting of North and South Dakota and Montana, form the Northern Great Plains Province. Combined with the chain of states that reaches from central Montana to northwestern Arizona and New Mexico—the Rocky Mountains Province—this expanse of coal is the youngest on the North American continent. The geologically undisturbed area of the Northern Great Plains Province is the nation's richest source of lignite. In the Rocky Mountain region, geochemical alteration has produced rich seams of bituminous coal, as well as the only major source of subbituminous coal in the United States. As the quality of coal becomes increasingly critical from an environmental point of view, coal from this region is becoming more valuable in spite of its generally lower rank. Many of the seams in this province have a very shallow overburden and are as much as 40 ft thick over huge areas. Also, the coal taken from these mines is much lower in sulfur content than are other coals, which many people believe more than compensates for its lower heating value. The wealth of coal in this region was largely unexploited until after World War II. Since then, the percentage of the total United States coal production taken from this region has increased annually and was about 29% in 1983 (31).

In the Pacific Coast Province, only Washington contains major coal reserves. Coal has been identified in western Oregon and in areas of California, but no production figures are given for those states in the current literature. Alaska is also included in this province and is a minor coal-producing state.

1.6 NAMING OF COAL SEAMS

Anyone who makes more than a cursory inquiry about coal will discover quickly that both the coal deposits and individual coal seams within a particular formation are named. This can be both a charming and a disarming discovery. First of all, because the same seam in a particular formation may extend across a large area—even across several states—different names may have been given to it in different locations. Second, no effort has been made to develop a uniform system of nomenclature or to adapt any other formal stratigraphic system to coal. Seams may thus be identified by a number, a town or river in the area, an Indian tribe, the name of a land owner on whose property a deposit was discovered, or even some early owner's pet name. Sometimes two types of names are used, usually a proper name and a number. Even when numbers are used, the system is not necessarily consistent. In Europe, seams are named or are numbered consecutively from the top down (34). However, in the United States the numbers increase from the deepest vein toward the surface. This pattern is not free from confusion. A vein deeper than Number 1 may be discovered at some later time. For instance, in Indiana, two formations of five seams are known below the Rock Island (Number 1) seam, which, unfortunately, is also known as the Litchfield seam and the Assumption seam. As already mentioned, the name of a seam may change as it crosses county and state lines. For example, the coal seam named Herrin (Number 6) in Illinois is just Herrin in Indiana, but it is 11 (Herrin), and just 11 across three counties in Kentucky. The use of people's names is fascinating. One seam in the Rhenish-Westphalian Basin in Germany has been known by 16 different names (34). In eastern Pennsylvania, the locality of a seam is often used as the trade name for the coal produced there. Lehigh, Scranton, Reading, Clearfield, and Somerset are all towns still producing coal. Big Mary and Blue Gem are less readily located in that state. McClearies Bluff, St. Meinrad, Paradise, Ice House, Upper Otter Creek, Mud River, and Elm Lick are examples from the midwest. In the Rocky Mountain Province and the Northern Great Plains Province, Raton, Belt Creek, Henryetta, Broken Arrow, Black Diamond, Snoqualmie (Numbers 3, 4, and 5), Beulah Gap, Nenevah, and Dead Man are some of the more colorful names. Texas uses only numbers; Wyoming uses large numbers (e.g., 24, 50, 82); and Alaska's coal remains unnamed altogether (35).

1.7 PROSPECTING, MINING, AND TRANSPORTATION

In presenting various working definitions used to describe coal resources (Section 1.4.1), the distinction was made between identified, hypothetical, and economically recoverable reserves of coal. Unless coal formations can be identified, mapped, mined, evaluated, and transported to a consumer, their potential is only of academic, rather than of practical, interest.

1.7.1 Prospecting

Both geologic and geographic influences must be considered in order to establish the possibility and the practicality of a mining operation. As pointed out in the beginning of this chapter, coal is considered a mineral, and therefore coal lands are treated in the same way as other mineral lands. This means that coal lands are under the jurisdiction of the Department of Interior. A number of regulations pertain to use of these lands, including ones requiring the formal acquisition of a lease or permit before exploiting coal deposits (36).

Prospecting for coal, as for any other material, involves two facets: the initial search, or exploration, to find coal; and the subsequent testing or evaluation of coal that is found. This testing includes the assessment of both the quantity and the quality of the coal. Exploration involves the implementation of geologic and engineering theory, but at the outset a search for coal should be conducted according to some general guidelines. The presence of shale, slate, or sandstone is compatible with coal deposition. Limestone is seldom associated with the presence of coal (37). Visual observations may reveal outcrops of coal or of dark and carbonaceous shale, which do suggest the presence of underlying coal. Sooty material called smut or blossoms is also a readily recognized indicator of the position of a seam. Drilling wells, or core boring, is the first step in the mapping or the geological survey of a coal formation. The need to drill to increasing or variable depths moving away from an outcrop indicates that a seam is not horizontal. Core samples may also reveal the presence of more than one seam within a single formation.

E. S. Moore wrote one of the two most heralded textbooks on coal. It is not possible here to do more than summarize his extensive analysis of structural features of coal seams (38). One particularly pertinent consideration that he describes is the potential effect of faulting on prospecting by core drilling. A fault is a break in the earth in which one side of the break is raised or lowered with respect to the other. In some cases displacement of the sides may be considerable. Faulting may result in either the appearance of an outcrop or its disappearance. It is more important that faulting may cause a seam to "double back" on itself, and a misplaced core sample will indicate erroneously the presence of two seams. Conversely, the core sample may miss the seam completely if it is taken at a point where there is a gap in the formation. Similar effects may result from less severe overlap of adjacent sections of the earth, called folding; the reader is directed to Moore's text for a more detailed description of these considerations. A number of techniques have been developed to correlate the profile of seams within a formation. This is particularly important if the formation dips and is composed of a number of seams, because it has been found that different types and ranks of coal can exist in the same formation. Extensive use is made of plant and animal fossils as a means of tracing coal beds. The prominent geologist R. Thiessen has shown that spores and other coalified organic matter tend to be unique from one seam to another (39). Another way to trace the continuity of a coal seam is to look for the presence of peculiar rock inclusions or to find similar concretions incorporated in a seam or in segments adjacent to a seam. Geologists also use the path of a water course to follow a seam line. One less common geologic phenomenon by which a seam may be mapped is the presence of reddened sandstone caused by previous (or ongoing) underground combustion. The resultant line of red cinders indicates the location of the bed. Other natural seam markers, such as a particular-colored mineral species, are also helpful in making correlations.

Some physical characteristics of coal seams can also prove valuable in exploration, such as fracture, streak, and cleat. Fracture, as the term implies, is the manner in which a coal breaks. The type of fracture can be related to the rank of a specimen. Bituminous and subbituminous coals break with cubicle fracture. Anthracite breaks with conchoidal fracture, leaving a concave surface. The streak of a mineral is the color exhibited when the powdered sample is rubbed on unglazed porcelain. For coal, the streak varies gradually with rank from yellow-brown for lignite, to black for anthracite. This characteristic is not generally useful in seam correlation, however. Cleats are a set of fractures that cut across a coal seam almost perpendicularly to the face of the seam. The face cleat is parallel to the face of the seam; the butt cleat is perpendicular to the face. Cleats will not cross a layer of clay or inorganic partings (narrow separations) between seams. Cleats are probably caused by gradual loss of moisture during coalification (40). The presence of cleats is extremely important in underground mining operations because improper mining techniques can result in explosions of the coal into the mine room. As long as a seam is undisturbed, the cracked seam is supported by the surrounding rocks. When the opening or broken area is exposed during mining, the pressure of the remaining mass of the seam may cause it to open suddenly with explosive force.

Testing to determine the thickness and extent of either an entire coal deposit or its individual seams involves the digging of channels, pits, or shafts for shallow seams and additional core sampling for deeply buried seams. Because coal deposits are seldom horizontal, the angle at which core holes intersect the seam can be used to calculate its dip. Less rigorous but satisfactory graphical methods have also been developed to determine the geographic extent and the dip of a seam. The simplest case involves finding the depth at various distances from the outcrop along the line of a seam that dips continuously from the outcrop. The depth at any distance from the outcrop is then directly proportional to the distance.

A similar method is used to find the depth at some point from the outcrop but not along the line of the seam. Such calculations as well as correlation charts are described in Moore's text but are beyond the scope of this book (41).

The tonnage available in a coal prospect is calculated in a unit known as a foot-acre and is directly related to the specific gravity of the coal (41). A foot-acre is the number of cubic feet per acre (43,560 ft^2). The specific gravity of any material is defined as the ratio of the mass of a given volume of the substance to the mass of the same volume of a standard (water at 4° C for solids and liquids) and is, therefore, a unitless number. A cubic foot of water at 4° C weighs 62.5 lb. The specific gravity and the weight of coal varies with its rank due to variations in its moisture and mineral content. For example, in one instance a cubic foot of a lump of anthracite weighed 57 lb, and the same volume of a block of anthracite weighed 92 lb (41). The average value of the specific gravity of bituminous coal is 1.30. A cubic foot of coal with an average specific gravity of 1.3 weighs 81.25 lb, and there are 24.3 ft^3 per short ton. If an acre completely underlain with 1 ft of coal could be 100% mined, it would yield 1793 tons of coal. It should be remembered, however, that while most seams are much thicker, the percentage actually mined is often substantially less than 100%.

1.7.2 Mining Operations

In 1929 it was forseen that advances in technology would make coal mining an art (42). Certainly, mining then was a simple matter of applying sufficient labor to the effort once the geologic, geographic, and economic considerations regarding the feasibility of the mining operation had been dealt with. Because of increased sophistication brought about by recent technology, modern coal mining may be considered simple because it requires less raw human physical exertion, but modern equipment is not particularly simple. Besides being an art, mining is now also an extremely complex business operation (43).

Basically, what has not changed in the last 50 or so years are the two principle mining methods: strip, or "open work," mining and underground, or "closed work," mining.

Strip mining is the simplest operation. The overburden is removed, and the coal is dug, loaded, and hauled away. Mining technology has brought about changes in the form of the tools used to carry out each of these tasks. Formerly, a hand-operated shovel removed the superficial earth. Horse-drawn scrapers and even motor-powered shovels have been replaced by a device called a dragline. A dragline is an excavating device, positioned adjacent to a seam, that removes the overburden in a large power shovel attached to the movable jib. The use of a dragline is similar to the use of a plow to make a furrow, except that with a dragline the object is to remove the overburden altogether. Forty years ago, a scoop capacity of 32 yd^3 was large. Today, shovel capacity has been extended to more than 200 yd^3 (44). Similarly, loading and hauling operations use the most advanced methods available. Strip mining is becoming increasingly more economical, more rapid, and safer than underground methods, but even with modern equipment, the character of the overburden is important and may limit the feasibility of strip mining. In general, the rate at which the overburden is removed is determined by whether it consists largely of loose soil and shale, or whether substantial rock beds require preliminary blasting. Because of the attractiveness of strip mining, even relatively thick layers of overburden are removed if the underlying coal seam is evaluated as warranting such a measure. At Pennsylvania's Mammoth seam, for instance, 125 ft of covering has been stripped in order to permit maximum access to the 100-ft thick seam of highly desirable coal. In western Kentucky, 90 ft of overburden has been removed over a seam only 2 ft thick. The coal is desirable because it has a very low ash content. Routinely, an overburden-to-seam thickness ratio of 20:1 is considered the upper limit.

By numbers, the advantages of strip mining surpass its disadvantages. Among its advantages are (1) more of the coal is removed; (2) less labor is required per ton produced, and the miners require less training and experience; (3) no timbering is required; (4) larger hauling units can be employed; (5) it is more rapid; (6) equipment can be moved to subsequent sites or sold; and (7) danger to miners in the form of poor mine ventilation and mine roof collapse is eliminated. The primary disadvantage from the production point of view is that weather conditions can limit or eliminate mining operations for lengthy intervals, causing substantial financial losses. The extensive scarring of the land caused by destruction of all the vegetation and soil is a disadvantage of considerable and growing concern. For the most part, with no externally applied pressure to do otherwise, strip mining sites have been abandoned with only minimal effort to restore the land. Recent public awareness has resulted in legislation requiring that various specific efforts be made by coal companies to reclaim depleted mine sites and restore them to conditions that will again permit plant growth and support animal life. This topic is the subject of ongoing debate and controversy.

Three types of mine openings are used to gain access to coal buried too deep to strip mine, and the method of choice depends largely on the geography of the land. Drift mining is used when a coal seam lies in a hillside. A drift is an opening drilled into the hillside head-on to the seam. An adit drilled at a sloping angle from the surface to the seam is called a slope access. This is commonly used in hilly areas or where the seam just exceeds reasonable strip mining limitations. When the deposit is deep and when the land is level, shaft mining is usually used. In all three cases, safety considerations are paramount, and laws govern the number and size of all access openings (45).

The oldest underground mining method is termed conventional, or room and pillar. This was the first and is the most complicated mining method ever used. It consists of working out rooms and leaving behind coal pillars to support the ceiling. In addition to the main ventilation and access shafts, a maze-like series of arteries is cut into the seam. The pattern of rooms and pillars is developed according to a definite, well-planned system. Laws govern the minimum size of the pillars and the maximum number of rooms allowed in a seam. Obviously, roof control is critical, and fatal cave-ins occur. Although considerable mechanical innovation has been applied to this method, the use of pillars of coal still results in the inability to remove 35–70% of the available coal. Some or all may be removed in a later operation, called robbing pillars, in which an intentional, controlled cave-in is caused. The average recovery of the coal using this two-stage process is 80%.

The longwall method, used for many years in Europe and becoming more common in the United States, involves the use of a mechanical self-advancing roof. Rather than cutting the coal face to leave pillars supporting the ceiling, movable steel beams are attached to a shearing mechanism. As the coal is shaved away and falls to a conveyor, the unit moves forward, allowing the unsupported ceiling behind it to collapse in a controlled and uniform manner. As much as 90% of the coal is recovered using this method. It is certainly a more rapid and simpler method than room and pillar. Compared to conventional methods, ventilation problems, especially from coal dust, are diminished and no "dusting"—coating of the rock face to control dust—is necessary. On the other hand, this system requires more experienced operators, it is more awkward and difficult to initiate, and it works best in less-valuable mines where the seams are thin (4½ feet or less) and where there is considerable waste rock (46).

A continuous miner is another excavating device used in underground mining. It has a front cutting head consisting of several continuously revolving chains carrying cutters or picks. The unit is mounted on caterpillar treads and uses a conveyor belt to move coal away from the face to shuttle cars following the unit. This device can cut an entry 12–20 ft wide and up to 8 ft high.

In addition to safety and land reclamation, three other coal mining problems require constant and frequently costly attention. Subsidence of the surface into mined-out deposits has already been alluded to. Although the collapse of the roof is a controlled process in longwalling operations, additional and unexpected surface subsidence can occur at any time above an abandoned mine. Considerable damage and loss of life can occur. Long-abandoned mine sites in Rhode Island have recently begun giving way under the burden of building that has gone on, in many cases, without knowledge of the abandoned mines below. It has renewed the interest of many people in the entire Narragansett Basin; large amounts of money have been invested to obtain core samples and to identify the expanse of old shafts.

Pollution of surface and underground water from acid mine drainage is another problem. Oxidation, principally of the minerals, results in the formation of sulfuric acid and discoloring yellow iron sulfates. Pyrite and marcasite are two forms of iron found in coal. Both have the formula FeS_2; they differ only in their crystalline form. In addition to normal or routine drainage from a mine while it is actively being used, the problem is compounded when acid water percolates to the surface as subsidence occurs.

Centralia, Pennsylvania, in the heart of the rich anthracite field, is currently being destroyed by what is perhaps the ultimate mine problem: underground fire. In this case, the fire has been burning for at least 21 years, after being started when a garbage dump was located and ignited on an outcrop. The fire burned underground for some time before it began to cause fissures in the surface that spewed a variety of noxious and toxic fumes into the city streets, sewers, and water lines. The extent of the fire is unknown, and many attempts to extinguish it have failed. What is known is that the damage to material possessions and the lives of the residents has been irreparable. In August 1983, about one-half of the residents agreed to abandon their homes in order for the federal government to attempt a massive containment project. The cost of the salvage operation is expected to exceed $100 million.

1.7.3 Transportation

Sometimes exceeding the cost of bringing coal out of a mine, the cost of transportation has a tremendous impact on the economics of coal utilization. The shipment of hundreds of millions of tons of coal annually is a major part of the transportation industry. The initial westward expansion of the nation's railroads was due in part to the need to bring coal to the new frontier for heating and for new industries. Railroads still carry about 65% of all the coal shipped within the United States and about 90% of all the coal being moved to seaports for export. Rail service is literally the lifeline of 98% of all western mines, considered captive by their dependence on rail service for shipping and for receiving materials (47). Barges, trucks, and conveyor systems are other methods sharing the rest of the market. A new development in coal transportation is the coal slurry pipeline. Still under study, this method remains controversial, and consequently there is only one commercial unit in operation. Actually, the concept of moving coal by pipeline is 100 years old. It was used in England before the turn of the century but was not used in the United

States until 1957, when the cost of rail transportation rose and made pipeline construction, operation, and maintenance costs competitive. A 273-mi pipeline now sends coal from the Black Mesa mine in Arizona to a generating facility in Nevada at the rate of 660 tons per hour. This pipeline has the capacity to transport 48 million tons of coal per year but has transported less than 1% of the coal transported from mine sites annually. While the use of less expensive and more expendable liquefied gases, such as CO_2, are being studied as potential media into which finely ground coal can be dispersed and pumped, water is used currently. Regularly spaced pumping boosters maintain the flow of the slurry in the pipe, and the coal is easily separated from the water at the terminals. Pipeline conveyance is particularly attractive in the west where coal must be shipped over substantial distances to reach large industrial consumers, but in addition to the potential threat presented principally to railroads by such a system, the use and control of such huge quantities of water (equal to the weight of coal) in a notoriously water-poor region is seriously disputed (48). As the use of western coal becomes increasingly more prevalent, the real possibility of west-to-east shipment by this method increases the seriousness of the issues.

In an effort to stay abreast of the competition for transportation service, the railroads have employed the concept of the unit train with great success. A unit train carries only a single payload between set points. Several engines are used to pull more than 100 cars, which can weigh as much as 100 tons each. The system has been designed so that loading and unloading can be done without stopping the train. This has lead to a several-fold increase in the rate at which coal can be moved from mine site to consumer (49).

The trucking industry is used for relatively short-haul needs, primarily from the mine to points linked to other transportation systems. Although of limited use, trucks carry as much as 13% of the coal moved in the United States (50).

Except perhaps for small-scale conveyor belt systems used at mine sites or between a mine and a mine-mouth consumer, transportation of coal on waterways is the least expensive. Recent estimates are that 11% of all coal is moved by barge (51). Like trains, large barges with a capacity of more than 1,500 tons are linked together, and loaded and unloaded by automatic tipple systems. Of course, water transportation services all coal being exported. Loading systems at the nation's dozen ports can transfer 100,000 tons of coal per hour to waiting ships (52).

The time element involved in coal transportation, especially with regard to the considerable income received by exporting companies, is not just an exercise in ever-increasing speed. Having contracted to provide a specific tonnage of coal whose quality is carefully defined and monitored, a mining company may lose thousands of dollars per shipment, and possibly damage its reputation, if the product delivered no longer meets the initially stipulated specifications. Because coal is an organic material containing reaction-promoting moisture and inorganic components, the longer the time lapse from mine to destination, the greater the possibility—and probability—that the quality of the coal will diminish. This physical and chemical loss of quality is called decrepitation. It occurs at an exponential rate and may result in total destruction of a quantity of coal by spontaneous combustion. In an effort to minimize or eliminate such processes during all transportation and storage periods, research is being carried out to determine what makes coal susceptible to decrepitation.

From prospecting activities through final delivery of coal to the consumer, the maintenance of the delicate balance among all aspects of the coal mining industry is a complex accomplishment that is frequently unrecognized and unappreciated by the general population.

1.8 COAL PRODUCTION AND CONSUMPTION IN THE UNITED STATES

Before 1964, no state west of the Mississippi River produced as much as 1% of the nation's total coal. Together, the 15 western coal-producing states produced only 5% of the total (32). The level of production from this region has increased dramatically since the mid-1960s. In 1970, the production value had risen to 11%, and in 1983 the west accounted for nearly 30% of the national coal production (53).

1.8.1 Nationwide Coal Production

From the end of World War II until 1973, coal lost more than 70% of its domestic market. In 1945, coal was the primary source supplying 50% of the United States energy needs; by 1974, that figure had dropped to 18%. Conversion from coal-fired steam to diesel engines decreased the railroads' demand for coal 600-fold, from 63 million short tons annually to 0.1 million short tons by 1973. Replacement of coal by oil, natural gas, and electricity for residential and commercial heat caused a 10-fold decrease in demand, from 115 million short tons to 11 million short tons during the same period (54).

The energy crisis of the mid-1970s prompted a major reappraisal of the nation's energy resources and the nature of our dependency on other countries. Coal reemerged as a significant part of the new energy plan.

Table 1.4 shows that from 1976 through 1980, when the world crisis in oil was still of paramount concern in this country, coal production increased steadily, from a net 12 million tons from 1976 to 1977 to a net 49 million tons from 1979 to 1980. Beginning in 1981, however, a decline began in coal production for the second time in recent history. Production for 1980 and 1982 is about the same, 830 and 838 million tons annually, but by 1983, coal production had fallen to 776 million tons, a decrease of 61 million tons. The reasons for the declining productivity during those years can be traced principally to increasingly restrictive environmental constraints on the allow-

TABLE 1.4 Annual Coal Production in the United States, 1976–1984

	1976	1977	1978	1979	1980	1981 [b]	1982	1983	1984 [c]
Production total [a]	684,913	697,205	670,164	781,134	829,700	823,775	838,112	776,635	895,900
Consumption total	603,790	625,291	625,225	680,524	702,729	732,627	706,911	736,672	791,000
Exports	60,021	54,312	40,714	66,042	91,742	112,541	106,277	77,772	82,000
Imports	1,203	1,647	2,953	2,059	1,194	1,043	742	1,271	1,300

Source: Reference 55.

[a] All values are in thousands of short tons.

[b] Production was affected adversely by a labor strike during the first quarter of 1981.

[c] Ref. 57, p. ix.

able levels of stack emissions and to the diminished interest and activity on the part of government and industry in investing in the long-range development of synthetic fuels from coal. Also, mine safety laws became more stringent, and capital outlay costs for mining operations have caused some decrease in mining activity. Total production did increase in 1984, but slowed again through most of 1985. Geographical productivity continued to shift steadily from the East to the West. Although approximately 60% of the mining occurred in the Eastern Province in 1979 and 71% of the nation's coal came from the states east of the Mississippi River in that year, the figures dropped, in 1983 and 1984, to less than 65% for that region. This decrease is equivalent to about 184 million tons of coal. One of the critical limitations of the utility of eastern coals that is no longer compensated for by their high heat content is their correspondingly high mineral content, particularly their sulfur content. We have already said that the lower sulfur content of western coals somewhat offsets their lower heating value. However, other factors must be taken into account to assess the overall potential of low-rank coal. Coal seams in the west are generally not buried under much overburden. This makes easier and lower-cost strip mining possible.

The exporting of coal, which was begun near the end of the last century, is also a major part of our coal economy. There was optimism that with the high levels of production at the beginning of this decade, exports would also increase. In 1980, exports did increase 40% over the previous year; nearly 92 million short tons of all ranks of coal were exported. The figure rose to 113 million short tons in 1981 but declined by about 6% in 1982. In 1984, an additional decline of 28% occurred, and only 82 million short tons were exported (Table 1.4). In 1983, Latin American countries purchased large quantities of United States coal, and Canada alone invested $1 billion in our coal. At the same time, however, other countries had cut back on the amount of coal they imported from the United States. By the middle of 1984, exports were down 50% from 1983. The overall world economy was faltering; our own national debt, the value of the dollar abroad, and the value of international currency were such that the acquisition of United States coal had become unattractive from an economic point of view. Poland, South Africa, and Australia are our major competitors for the export market. Restabilization of labor within other coal-producing countries and renewed efforts to develop their own energy resources have all combined to spur production of coal abroad and to diminish the need to buy coal from the United States.

There was optimism, however, that total coal production will not continue to sag. The primary reason for that optimism is that data suggest that demand for electricity will force electric utilities to expand their operations. Uncertain oil prices are expected to cause utilities to convert back to coal-fired boilers. The fluctuating oil price market is also expected to aid our coal export trade favorably in spite of the uncertain world economic picture. Table 1.5 gives recent projections of total coal production and the projected demand for coal by the major coal consumers. These figures show that increased demand for

TABLE 1.5 Projected Production and Consumption of Coal in the United States, 1985–1995

	1985	1990	1995
Production			
East	582	650	731
West	317	407	490
Total	899	1,057	1,221
Consumption			
Electric utilities	709	805	951
Industrial			
Metallurgical	46	51	49
Other	77	89	98
Residential, commercial	8	7	7
Synthesis	5	6	6

Source: Reference 57, p. 24.
Note: All values are in millions of short tons.

coal will be met by both eastern and western mining operations in about the same proportion as in 1984. There is a growing interest in developing the use of lignite, but this is not expected to make a substantial difference in production from mines in the west in this decade.

1.8.2 Production Classified by Rank and Mining Method

Table 1.6 shows recent data for surface and underground mining. In the west, only five states employ underground methods, and only 6% of all coal mined underground is taken from this region. Figures released for 1984 showed a 15% overall increase in coal production from the previous year, but the ratio of surface to underground mining operations nationwide remained about the same. The production figures for 1984 are somewhat higher than were predicted and are higher than preliminary figures for 1985 in part because extensive stockpiles were built up early in 1984 in the event of a nationwide strike by miners later in the year. It is also interesting that three states—Pennsylvania, West Virginia, and Kentucky—still produce nearly 41% of the nation's total coal and 90% of all the bituminous coal mined nationwide. Table 1.6 also shows coal production by rank for each principal coal-producing state. The data emphasize that the major portion of the coal mined is taken from deposits east of the Mississippi River but also show clearly how drastic the effect will be on eastern states as interest in low-rank coal increases. In Wyoming, Texas, and North Dakota, virtually all of the coal produced is low-rank; Pennsylvania, West Virginia, and Kentucky produce primarily bituminous coal.

TABLE 1.6 United States Coal Production by Coal-Producing State, Type of Mining, and Coal Rank, 1983 and 1984

	1984					
	Type of Mining				Total	
	Underground		Surface			
Coal-Producing State	Production	Percent Change vs. Prior Year	Production	Percent Change vs. Prior Year	Production	Percent Change vs. Prior Year
Alabama	13,163	21.21	13,855	8.54	27,019	14.37
Alaska	—	—	859	9.31	859	9.31
Arizona	—	—	11,522	1.03	11,522	1.03
Arkansas	—	—	60	−3.42	60	−3.42
Colorado	6,399	14.65	11,547	3.73	17,946	7.38
Georgia	—	—	125	−32.56	125	−32.56
Illinois	38,498	20.92	25,259	1.01	63,756	12.16
Indiana	2,242	25.78	35,297	17.64	37,540	18.09
Iowa	168	—	358	−2.71	526	43.01
Kansas	—	—	1,328	5.13	1,328	5.13
Kentucky	78,240	20.69	78,535	22.96	156,774	21.82
Eastern	60,925	24.31	53,720	21.57	114,645	23.01
Western	17,315	9.47	24,815	26.10	42,130	18.69
Maryland	2,206	33.98	1,850	24.72	4,056	29.59
Missouri	—	—	6,724	35.13	6,724	35.13
Montana	—	—	33,000	14.09	33,000	14.09
New Mexico	552	436.29	20,717	1.99	21,269	4.19
North Dakota	—	—	22,108	15.26	22,108	15.26
Ohio	14,102	30.31	25,013	9.90	39,115	16.47
Oklahoma	—	—	4,598	25.48	4,598	25.48
Pennsylvania	37,253	7.05	39,411	15.71	76,664	11.33
Anthracite	399	32.38	3,470	.60	3,869	3.16
Bituminous	36,854	6.83	35,941	17.41	72,795	11.80
Tennessee	5,196	19.24	2,014	−8.75	7,211	9.83
Texas	—	—	41,145	5.64	41,145	5.64
Utah	12,323	4.82	—	—	12,323	4.82
Virginia	32,627	21.84	7,279	−5.93	39,906	15.61
Washington	—	—	3,863	−.54	3,863	−.54
West Virginia	105,839	15.14	24,456	9.32	130,295	14.00
Wyoming	1,345	7.60	129,569	16.77	130,914	16.67
East of the Miss. River	329,366	17.79	253,094	13.78	582,460	16.01
West of the Miss. River	20,788	11.22	287,397	12.32	308,184	12.25
U.S. Total	350,154	17.38	540,490	13.00	890,644	14.68

TABLE 1.6 (Continued) United States Coal Production by Coal-Producing State, Type of Mining, and Coal Rank, 1983 and 1984

1983						
Production						
Type of Mining			Coal Rank, 1984			
Underground	Surface	Total	Bituminous Production	Subbituminous Production	Lignite Production	Anthracite Production
10,860	12,765	23,625	27,019	—	—	—
—	786	786	—	859	—	—
—	11,404	11,404	11,522	—	—	—
—	62	62	60	—	—	—
5,582	11,131	16,713	11,367	6,579	—	—
—	185	185	125	—	—	—
31,838	25,007	56,845	63,756	—	—	—
1,783	30,005	31,788	37,540	—	—	—
—	368	368	526	—	—	—
—	1,263	1,263	1,328	—	—	—
64,826	63,868	128,694	156,774	—	—	—
49,009	44,190	93,199	114,645	—	—	—
15,817	19,678	35,494	42,130	—	—	—
1,647	1,483	3,130	4,056	—	—	—
—	4,976	4,976	6,724	—	—	—
—	28,924	28,924	—	32,771	229	—
103	20,312	20,415	12,377	8,893	—	—
—	19,181	19,181	—	—	22,108	—
10,822	22,760	33,582	39,115	—	—	—
—	3,664	3,664	4,598	—	—	—
34,799	34,061	68,860	72,795	—	—	3,869
301	3,449	3,750				
34,498	30,612	65,109				
4,358	2,208	6,565	7,211	—	—	—
—	38,947	38,947	416	—	40,729	—
			12,323	—	—	—
11,756	—	11,756	39,906	—	—	—
26,779	7,738	34,517				
—	3,884	3,884	—	3,863	—	—
91,918	22,371	114,290	130,295	—	—	—
1,250	110,963	112,213	4,688	126,226	—	—
279,628	222,451	502,079	578,591	—	—	3,869
18,691	255,865	274,556	65,928	179,190	63,067	—
298,320	478,315	776,635	644,519	179,190	63,067	3,869

Source: Reference 57 pp. 14, 24 when updated.
Note: Excludes silt, culm, refuse bank, slurry dam, and dredge production except for Pennsylvania Anthracite. Excludes mines producing less than 10,000 short tons of coal during the year. Total may not equal sum of components because of independent rounding. Data are in thousands of short tons.

1.9 VALUATION AND USES OF COAL

Several factors must be taken into account when assessing the overall projected value of a particular coal prospect. Among them are (1) the proximity of the mine to the prospective consumers; (2) the availability and prevailing cost of transportation and labor; (3) the competition, or the amount of equivalent products, from all other sources in the area; (4) the quality of the coal; (5) the quantity of the coal; (6) the accessibility of the seams; and (7) ancillary considerations, such as the availability and cost of water and equipment and overhead expenses, such as land taxes and the projected cost of land reclamation. The emphasis given to any one or a combination of these criteria must be decided separately in each situation. Certainly, one factor that is a liability can be offset by another, highly favorable factor. In addition, the knowledge that coal is available may actually stimulate a sluggish market or meager transportation facilities. All of the same aspects of the economy that affect other corporations apply to the coal industry, and the assessment of a dollar value per acre for coal lands is subject to regular fluctuations.

The primary consumers of coal since 1980 have been electric power producers. The greatest increase in demand for coal was made by the utilities, which burned more than 569 million short tons of coal in 1980, 70% of the total produced that year. In 1985, 550 million short tons (76%) were shipped to electric utility companies. It is projected that by 1995, electric utilities will use 916 million tons of coal annually (Table 1.5). The steel industry is the second-largest coal consumer. In 1980, coke production used 67 million short tons of coal. The United States steel industry has been depressed, however, and in 1983, only 37 million tons of coal were consumed by the steel industry (55). Data in Table 1.5 do show, however, that some improvement in the steel industry is expected during the next decade.

Historically, blacksmiths were the principal users of coal. In colonial America, home heating and cooking were done with wood. Even early blast furnaces used charcoal instead of coal because it was abundant and inexpensive. Anthracite became popular as a home heating fuel after the Revolutionary War when it was shown that it was clean to handle and burned with almost no smoke. The use of anthracite increased rapidly until after World War II when natural gas and oil became cheaper. Anthracite has a very limited potential for industrial use; even the renewed interest in coal-burning stoves that has occurred in the last few years is not likely to cause a dramatic surge in the demand for anthracite. The primary barrier to its widespread use as a domestic fuel is the limited geographical area in which it is found. Transportation costs to areas outside the northeast are prohibitively expensive.

Coal, in the form of coke, spurred the American industrial revolution when it replaced charcoal as a furnace fuel. The formation of coke from coal is analogous to the production of charcoal from wood. Coke is produced when certain bituminous coals—coking coals—are heated in the absence of air.

The volatile components are driven off and leave a fused carbon residue. Another name for the coking process is carbonization. Carbonization at low temperatures (600–700° C) is used primarily to produce flammable gases and a solid residue similar to charcoal called semicoke. Higher temperatures (700–950° C) are used to produce a true coke. Metallurgical coke is prepared from lower-grade coal, while coke suitable for use in a foundry requires a high-grade coal, that is, a coal that is low in sulfur and other mineral matter. Most modern methods of coke production use mixtures, or blends of several ranks and grades of coal that will yield a product having the desired quality. Carbonization is a complex physical and chemical process, treated in detail in Chapter 4.

In conjunction with the development of coking processes, the abundant supply of coal for steam engines initiated the expansion of railroads across the United States at the end of the nineteenth century; railroads in turn spread the use of and demand for coal by giving mobility to the population and to industrial centers.

At the beginning of this century, the advent of coal-fired steam generators to produce electricity marked another milestone for coal utilization, but "King Coal" did not reign for long. The Depression years (1929–1939), the discovery of abundant and more accessible oil and natural gas, and technological advancements that made it possible to produce more energy with less coal all combined to alter drastically the course of coal production and consumption from what had been predicted before World War I (58).

The chemical industry uses coal as a primary raw material for a wide variety of products ranging from antiseptics and other pharmaceuticals to explosives. Dyes, perfumes, fertilizers, synthetic fibers, solvents, coatings, and ink are all prepared from coke or the by-products of coke manufacture. Coal tar obtained from low-rank coal is a source of a crude oil. Asphalt and roofing tars are side products. Lignite is a source of montan wax, the material once used to make phonograph records, wire insulation, and shoe polish (59).

The market demand for coal as a source of many of these items is being severely threatened by the use of synthetic materials, especially plastics and other polymers. It seems that the future of coal lies exclusively in its potential as a source of energy. Laboratories throughout the country are investigating ways of economically converting coal into a feedstock material from which the fuels whose production now relies on petroleum can be made. This is reflected in Table 1.5. In compiling the data in Table 1.5, the use of coal for synthetic coal-derived fuels was considered in projecting the demand for coal for the first time in history. In 1982, the United States Synthetic Fuels Corporation (SFC) was created to give financial support to selected projects designed by various industries to develop practical coal-to-fuel processes. Initially 14 projects were selected by the SFC for funding. Many of these projects are designed to produce gases from coal that can be readily used in synthesis reactions. The production of ammonia, methyl alcohol, and gasoline are three other goals of SFC-funded projects. The chemistry and technology involved in coal gasification and the subsequent synthesis of new products are presented in Chapter 4.

The Great Plains Coal Gasification Project was expected to begin production in late 1984. The facility will use as much as 14,000 tons of lignite per day and has been designed to produce daily 138 million ft^3 of synthesis gas, 93 tons of ammonia, 88 tons of elemental sulfur (a raw material for sulfuric acid, the chemical produced in largest volume in the nation), and 200 million ft^3 of carbon dioxide (60). The operation of similar plants across the country would have a dramatic effect on the coal industry. It remains to be seen, of course, to what extent these projects will be able to meet the original expectations.

1.10 CHARACTERISTIC PROPERTIES OF COAL

It was shown earlier that neither the physical nor the chemical separation between ranks of coal from lignite to anthracite is sharp; instead, a range of experimental values is used to designate the individual ranks of coal.

Due to the worldwide occurrence of coal deposits, the numerous varieties of coal that are known, and the many uses of coal, a number of coal classification systems have been developed. These systems often are based on characteristics of domestic coal without reference to coal in other countries. One result of this plurality is that terms describing similar or identical coals are not used uniformly in these various systems. In the United States as well as abroad, systems for classifying coal are now based primarily on characteristic properties determined by laboratory methods. The chemical analysis of coal, however, as it is carried out in commercial laboratories, is made on representative gross samples and not on the individual components. It is known that such components also have identifiable characteristics. This matter is discussed in more detail in Section 1.10.1 and in Chapter 2. To eliminate confusion in international trade and to facilitate the exchange of technical and scientific information related to coal utilization and research, an attempt has been made to develop an international system for classifying coal on the basis of standard chemical analyses. It is not desirable to present here a detailed description of classification schemes used outside the United States, but some comparisons with European systems will be included. Chapter 2 gives a detailed description of classification systems based primarily on physical measurements.

Many methods of coal analysis are empirical and involve strict adherence to specified conditions, such as sample and particle size, temperature, time, and rate of heating. Some of the names for classes of coal were established before the development of uniform analytical methods and frequently by more than one organization. The United States Geological Survey (USGS) and the United States Bureau of Mines (USBOM) are two such independent agencies that have studied coal without mutually agreeing on consistent descriptive criteria. For analytical work, the establishment of specifications that are recognized as standards and are supported by authoritative organizations is essential. In the United States, the American Society for Testing and Materials (ASTM) Committee D-5 has the responsibility for developing standard procedures for coal analysis. The discussion of the analysis of coal in Chapter 5 pertains primarily to the ASTM standard methods.

1.10.1 Ranks and Varieties of Coal

True brown coal is not common in the United States, but it is found throughout Europe, Canada, and Australia. It is typically called brown lignite or brown coal. Based on various analyses, the rank of some brown coal can be as high as subbituminous. The USGS has traditionally reserved the term *lignite* for the low-grade brown coal that splits and crumbles on drying, while the ASTM considers all low-rank brown and black coal to be lignite (61).

Subbituminous coal is the USGS designation of rank that includes a broad range of coals from a glossy black coal to compact "black lignite"; that is, it does not include material that is brown. On a physical basis, the demarcation between subbituminous and bituminous coal is the manner by which it fractures. Subbituminous coal breaks into slabs with irregular conchoidal fracture; bituminous coal has a characteristic cubical fracture. Bituminous coal also has a slightly higher specific gravity (62). Based on chemical analysis, increasing ranks of subbituminous coal are designated C, B, and A according to increasing heat content, as shown in Figure 1.2.

Bituminous coal generally burns with a yellow flame and emits a pungent odor. It is usually laminated, or banded, and the luster of different bands varies. It has been found that the bands consist of different chemical components and appear to have different chemical and physical properties (see Chapter 2). These differences exemplify the heterogeneous nature of coal, and the study of them contributes to a better understanding of the nature and utilization of coal. The name *bituminous* does not indicate that a coal necessarily contains bitumen, which is associated with petroleum. Because of the possible ambiguity of *bituminous,* some people prefer the term *humic coal.* Since bituminous coal may be either banded or nonbanded, and can include splint and semisplint coal, cannel coal and boghead coal, the use of *humic coal* is also inaccurate. Splint coal is a subdivision of banded coal with very fine bands, or microbands, and a dull luster. Cannel and boghead coal are two types of nonbanded coal. Cannel coal is composed predominantly of plant spores, ignites easily, and has a bright, blue, candlelike flame. It crackles while burning and is sometimes also referred to as parrot coal. It does not form coke when heated (it simply crumbles), but because it does have a high volatile content it is considered a gas coal. Boghead coals are distinguished from cannel coals largely on the evidence that they are composed of algae rather than spores. Boghead coal also releases a large quantity of illuminating gas when burned, but compared to cannel coal, it contains a higher percentage of hydrogen and mineral matter (11% and 20%, respectively, versus about 5% hydrogen and 5–13% ash in a typical bituminous coal).

As with subbituminous coal, there are subgroups of bituminous rank: high-volatile C, B, and A; medium-volatile; and low-volatile. The high-volatile C and B groups are assigned on the basis of increasing heat content; the high-volatile A group is characterized by both fixed carbon and Btu value. Medium-volatile bituminous is distinguished from low-volatile bituminous on the basis of fixed carbon content.

Another very important distinction within the bituminous rank is the coking or noncoking character of the coal. When heated sufficiently, coking coal softens and becomes a paste, a plastic mass; above the plastic temperature, it loses its volatile components and the hard, cellular coke remains. Noncoking coal may be caking coal (i.e., it may swell and soften), but the loss of volatile matter leaves a residue that does not hold its form under even mild stress and is not considered a useful coke. In general, medium-volatile bituminous coal is the best coking coal. Coking ability must be determined empirically. D. White found that the ratio of hydrogen-to-oxygen correlated well with coking ability. If the percent H:O ratio is .55–.59 or higher, subsequent tests of the coal consistently show that a good coke will form (63). Laboratory methods that are rather lengthy to carry out are described by ASTM and others (63,64). A rapid and nondestructive test of coking ability is made by rubbing a sample on a mortar. Coking coal clings to the sides, while noncoking coal does not.

Semianthracite coal ranks between low-volatile bituminous and anthracite. Numerous cleats, probably caused by rapid dehydration during its formation, cause this coal to be quite friable. Although it is a hard coal, it does contain a quantity of volatile matter, which causes it to ignite more easily than anthracite.

Anthracite is true hard coal and is commonly called stove coal. It is a clean-burning coal and is clean to handle as well, unlike bituminous coal. The heat content of anthracite is less than that of either semianthracite or low-volatile bituminous and less than that of some medium-volatile bituminous coals. It has few readily combustible constituents, but it is a desirable fuel because it burns longer. The term *hard coal* can be related to the standard Mohs' scale of hardness used by mineralogists. On the Mohs' scale, 2 corresponds to gypsum, 3 to calcite (crystalline $CaCO_3$, as in marble), and 4 to fluorite (CaF_2). The highest value is 10, the hardness of diamonds. Anthracite has the highest value of all coal on the Mohs' scale: 2.75–3. Although less desirable than many semiprecious gemstones, anthracite can be cut and polished. The specific gravity of anthracite is high also and varies between 1.5 and 1.7.

A particularly physically "pretty" variety of anthracite is called peacock coal. It displays an array of colors similar to that of a layer of oil on water. Some believe that the colors actually are due to a layer of oil or water-deposited iron oxide or sulfur dioxide. The most common explanation for the phenomenon is that it is caused by irridescent pieces of iron oxide formed by the oxidation of pyrite on the surface of the coal. No particular importance has been associated with this coal.

1.10.2 Systems of Chemical Analysis of Coal

It was shown in Section 1.10.2 that a number of chemical and physical properties of coal, or a combination of properties, are used to classify coals by rank. The best system is primarily a matter of utility. The best-known systems are based on two series of chemical analyses: proximate analysis and ultimate analysis. Proximate analysis is defined by the ASTM as the determination of moisture, volatile matter, and ash and the calculation of fixed carbon by difference (64). Ultimate analysis concerns the elemental content of coal. It includes the direct determination of the percentage of carbon, hydrogen, sulfur, nitrogen, and ash in the sample and the estimation of its oxygen content by difference (65).

Historically, ultimate analysis, rather than proximate analysis, was used first to designate coal rank. A French geologist, V. Regnault, introduced this system in 1837. It was modified in 1874 and again in 1911 (66). Regnault correlated the elemental composition and properties of the coal with its rank and included wood and peat as well as coal. Later modifications of his original work established limits for carbon, hydrogen, and volatile matter content for each rank and included the determination of heat content and a description of the coke product for each coal. The modified Regnault analysis included bituminous and anthracite coal only (66).

In the United States, P. Frazer used a system based on the ratio of fixed carbon to volatile matter (fuel ratio). This method was useful but did not include low-rank coals (67). Frazer introduced the method in 1877, and although it was modified several times in subsequent years, it never gained widespread popularity.

M. R. Campbell, an American geologist associated with the USGS, recognized that the determination of the fuel ratio and the simple determination of carbon content were not unambiguous methods of establishing rank. Likewise, he recognized that heat content was not a satisfactory parameter and that hydrogen content was of no use as a means of classifying coal. He preferred instead the ratio of carbon to hydrogen as a satisfactory method of classification. Campbell later modified his method and adopted the graphical schemes shown in Figure 1.2. This work of Campbell and the later work of S. W. Parr and G. H. Ashley led to the recognition that proximate analysis, rather than ultimate analysis, could be applied more readily to the task of chemically classifying coal.

Parr's classification scheme used the ratio of calorific value to volatile matter. He believed, correctly, that some material in coal will volatilize without contributing to the heating value of the coal. This process has already been alluded to in describing the significance of moisture in coal (Section 1.3.3). In addition to moisture, Parr also identified some mineral matter and sulfur as non–heat-producing materials and was able to develop empirical factors to compensate for the ash and sulfur content (68) (see Chapter 5).

Ashley's system addresses the natural moisture content of coal but correlates rank exclusively to the amount of fixed carbon (69). The classification system ultimately agreed upon by both the ASTM and the USBOM actually combined features of schemes proposed by Frazer, Parr, and Ashley.

Since the beginning of this century, several more complicated systems for classifying coal have been developed, principally for specialized applications. The texts by the English chemist W. Francis and the American geologist E. Moore give thorough discussions of the most prominent of these alternative systems. While each one has made valuable contributions to various aspects of coal analysis and classification, they have not achieved the widespread acceptance that the ASTM system has. Similarly, the National Coal Board (NCB) in the United Kingdom and the Economic Commission for Europe (ECE) independently developed classification schemes that have become incorporated to varying degrees in the system preferred by the British Standards Institute and the International Coal Classification system (70).

It is important to realize that up to this point we have been referring to both the chemical analysis of coal that has been carefully sampled and statistically treated and to the gross physical descriptions of hand samples of coal. For the most part, these two approaches have two entirely different purposes. The chemical analysis of coal serves to characterize the present state of the coal, to assign its rank and grade, and to provide a preliminary indication of how well suited the coal is for a particular end use, for instance, gasification or coking.

The physical descriptions and classifications of coal, on the other hand, are generally only the first step in a much more thorough study of the *past* history of the coal, its original composition and the environment or environments in which it existed in previous geologic time. This past history is revealed by careful and detailed microscopic analysis of the physical character of the coal. By combining data from the microscopic analyses and the chemical analyses of a coal sample, geologists and chemists can make correlations from which they are able to make projections regarding the utilization of a particular coal that are far more useful than relying on one descriptive approach alone.

SUMMARY

Although coal has been known and used by mankind for centuries, it has never been explicitly characterized either physically or chemically. In the current era sophisticated science and technology have finally made it possible to envision achieving a complete and an accurate knowledge about the nature of this complex and mysterious substance. This chapter has established a solid foundation from which a study of coal *science* can proceed. It provides a description of an array of historical, geological and commercial aspects of coal viewed principally as a commodity; it introduces the reader to the language used by professional coal scientists, and it provides the preface to the study of coal petrology, coal structure and analysis, and the utilization of coal which the subsequent chapters treat in detail.

We now have a thorough overview of the nature of coal, where it is found, and how it is obtained and classified. We are now prepared to delve into the origins of coal in geologic history in order to understand more fully those features of coal that are not apparent to the eye and are obliterated by chemical analysis but are nonetheless of ultimate importance to coal science.

REFERENCES

1. O. Stutzer, *Geology of Coal,* A.C. Noé, trans., University of Chicago Press, Chicago, 1940, p. 127.
2. E.S. Moore, *Coal: Its Properties, Analysis, Classification, Geology, Extraction, Uses and Distribution,* 2d ed., Wiley, New York, 1940, p. 1.
3. K. Lindberg and B. Provorse, *Coal: A Contemporary Energy Story,* R. Conte, ed., Scribe and *Coal Age,* New York, 1977, p. 18.
4. Stutzer, p. 88.
5. Moore, p. 2.
6. Lindberg and Provorse, p. 22.
7. Moore, p. 2; Lindberg and Provorse, p. 28.
8. *Coal: 1985 and Beyond,* Pergamon, London, 1978, p. 63.
9. G.J. Jansen, "The Petrography of Western Coals," *Proceedings of the Second Symposium on the Geology of Rocky Mountain Coal,* H.E. Hodgson, ed., Colorado Geological Survey, Denver, 1978, p. 181.
10. L. Petrakis and D.W. Grandy, "Coal Analysis, Characterization and Petrography," *J. Chem. Ed., 57,* 689–694 (1980).
11. Moore, p. 3.
12. T.A. Hendricks, "The Origin of Coal," in H.H. Lowry, ed., *Chemistry of Coal Utilization,* vol. 1, Wiley, New York, 1945, p. 2; W. Francis, *Coal: Its Formation and Composition,* Edward Arnold, London, 1961, p. 1.
13. Moore, pp. 2, 3; Francis, pp. 287–290, 452.
14. P. Averitt, *Coal Resources of the United States, January 1974,* Geological Survey Bulletin no. 1412, document no. I19.3:1412, United States Geological Survey, Washington, 1975, p. 7.
15. W.T. Thom, Jr., *Petroleum and Coal: The Keys to the Future,* Princeton University Press, Princeton, NJ, 1929, p. 87; Lindberg and Provorse, p. 16.
16. Averitt, p. 5.
17. Lindberg and Provorse, p. 16; D.N. Cargo and B.F. Mallory, *Man and His Geologic Environment,* Addison-Wesley, Reading, MA, 1974, p. 216; D.J. Cuff and W.J. Young, *The United States Energy Atlas,* Free Press, New York, 1980, p. 17; Moore, p. 228.
18. Cuff and Young, p. 17; Lindberg and Provorse, p. 16.
19. Thom, p. 87; Moore, p. 264.
20. Moore, p. 263.
21. Averitt, pp. 105–106.

22. Averitt, pp. 14–15.
23. *1984 Annual Outlook for U.S. Coal,* Energy Information Administration Report no. DOE/EIA-0333(83), Washington, 1983, pp. 25, 31–32.
24. C.G. Groat, "Coal," *Amer. Ass'n Pet. Geol., 65*(10), 2261 (1981); *Coal: 1985 and Beyond,* p. 65.
25. *Facts about Coal in the United States,* Environmental Policy Center, Washington, 1974, p. 2.
26. *The Story of Anthracite,* Hudson Coal, New York, 1932, p. 4.
27. *The Story of Anthracite,* p. 5.
28. H.J. Rose, "Classification of Coal," in H.H. Lowry, ed., *Chemistry of Coal Utilization,* vol. 1, Wiley, New York, 1945, p. 30.
29. Averitt, p. 17.
30. M.R. Campbell, *The Coal Fields of the United States,* United States Geological Survey Professional Paper no. 100-9, Washington, 1922, p. 11, cited in Thom, p. 47.
31. *Coal Distribution, January-June 1984,* Energy Information Administration Report no. E 3.11/7:984/2, Washington, 1984, p. 52.
32. *Facts about Coal in the United States,* p. 4; Lindberg and Provorse, p. 16.
33. Thom, p. 52; *Coal Distribution, January-June 1984,* p. 7.
34. Stutzer, p. 229.
35. *1981 Keystone Coal Industry Manual,* McGraw-Hill, New York, p. 493.
36. Moore, p. 251 ff.
37. Moore, p. 254.
38. Moore, pp. 228–250.
39. Moore, p. 255.
40. Stutzer, pp. 255–256.
41. Moore, p. 261.
42. Thom, p. 78.
43. Lindberg and Provorse, p. 44.
44. Moore, pp. 269–270.
45. Lindberg and Provorse, pp. 44 ff.
46. Moore, pp. 269–290.
47. Groat, p. 2261.
48. Lindberg and Provorse, pp. 138–140.
49. Lindberg and Provorse, pp. 133–138; V.D. Hanson, "The Unit Train Concept: A Revolutionary Idea," in M.E. Hawley, ed., *Coal, Part 2: Scientific and Technical Aspects,* Dowden, Hutchinson & Ross, Stroudsburg, PA, 1976, pp. 284–290.
50. *1984 Annual Outlook for U.S. Coal,* pp. 20–21.
51. Groat, p. 2261.
52. Lindberg and Provorse, p. 133.
53. Cuff and Young, p. 49.
54. Rose, p. 3.
55. *Monthly Energy Review, June, 1984,* Energy Information Administration Report no. DOE/EIA-0035(84/06), Washington, 1984, p. 73.
56. Rose, pp. x, 16.
57. *Annual Outlook for U.S. Coal 1985,* DOE/EIA-0118(84), Energy Information Administration, Washington, May, 1985.
58. Thom, pp. 99–101; Lindberg and Provorse, pp. 17–28.

59. Lindberg and Provorse, pp. 167, 181; Stutzer, pp. 128–130.
60. Rose, pp. 16, 30.
61. Moore, pp. 94–95.
62. Averitt, p. 21.
63. Moore, pp. 53–91.
64. *Annual Book of ASTM Standards,* vol. 05.05, section D3172-73, American Society for Testing and Materials, Philadelphia, 1984.
65. *Annual Book of ASTM Standards,* vol. 05.05, section D3176-74. American Society for Testing Materials, Philadelphia, 1984.
66. Francis, p. 363.
67. Moore, p. 113.
68. Moore, pp. 118–121.
69. Francis, p. 381; Moore, p. 128.
70. Francis, pp. 390 ff.

2

THE PETROLOGY AND PETROGRAPHY OF COAL

Of all the natural and synthetic materials known, coal is without a doubt the most unusual and probably the most complex. Statistics can show that the probability that eight to ten independent variables will attain a particular set of values, synchronously, is astoundingly small. And yet evidence supports the notion that such a set of circumstances did in fact occur in geologic history.

The goal of coal petrology and coal petrography is twofold: to look back and attempt to understand fully how coal was constituted, and to look forward and attempt to implement the most productive and beneficial use of all material classified as coal. Coal science, then, is necessarily a broad and a multifaceted discipline. It makes use of all the tools associated with zoology, botany, organic and inorganic chemistry, geology, and physics.

This chapter presents the best understanding of the best coal petrologists about when, where, and how coal was formed. It addresses the physical and chemical properties of coal, and it shows how these properties are used to classify and characterize coal.

A vital part of the nature of coal science is the historical development of ideas as well as the development of field and laboratory techniques. Both of these topics are treated in detail so that the reader can more fully appreciate the accomplishments that have been attained in this field of study.

2.1 THE NATURE OF COAL PETROLOGY AND PETROGRAPHY

Paleobotany is the area of specialization in botanical science that is concerned with the study of fossilized plants. It might be considered a science that is a bridge between the botanists' study of the development of the present plant world and the geologists' inquiry into the development in time past of the climate, topography, and rock strata of the earth. A coal paleobotanist is particularly concerned with the manner in which fossil plant material occurs in coal and in rocks associated with coal. In some respects the study of coal paleobotany focuses on the relationships among plants during the initial stages of coal formation.

Petrology is related to paleobotany but is a more general discipline; literally it means the study of rocks. Coal petrology concerns descriptive aspects of coal, such as the types of rock associations found in coal and the rock composition of coal. This information provides the physical foundations of the formation of coal.

Coal petrography, on the other hand, is the science that deals with the systematic characterization and classification of coal based on petrological data. Petrological and petrographic analyses include visual examination of hand samples as well as highly sophisticated microscopic analyses. Petrography has become indispensable for providing information that correlates the composition of coal and its technological behavior in coal utilization processes. In this text, petrology and petrography are addressed separately.

2.2 GEOLOGIC TIME: A COMPARISON OF THE PAST AND THE PRESENT

Geologists have divided the time frame encompassing the development of the earth into four eras. From earliest to most recent times, they are the Proterozoic, the Paleozoic, the Mesozoic, and the Cenozoic. These eras have each been divided into periods. Periods are divided into epochs, and epochs into ages. This scheme is illustrated in Table 2.1 and includes the estimated number of years associated with each division. The table also shows that coal has occurred in every geological period since the Silurian, more than 300 million years ago. Because the genesis of coal is intimately related to climatic conditions, it is not unexpected that the distribution of coal through the ages is not uniform. Small amounts of coal from the Devonian period have been found only in the Appalachian region of the United States, in Norway, and in the Soviet Union. The rate of formation of plant life increased rapidly, however, during the Pennsylvanian period. At that time, much of the land mass of the continents was located nearer the equator, and the climate over much of North America was tropical or subtropical. It is believed that areas containing shallow water were widespread and that marine life was abundant (2). These conditions fostered the growth of the first forests across much of North America. Vast deposits of coal were formed in a number of locations during this time. Coal from the Mississippian period is found in regious of Virginia, Oklahoma, Arkansas, and Alaska. Most coal deposition from the Mississippian period is associated with northern Europe, the Soviet Union, and Asia and less so with the North American continent. The coal found in the eastern United States originated during the Pennsylvanian and the Permian periods. Coal deposits from these periods are also found in South

TABLE 2.1 Description of Geologic Time

Era	Period	Time Since Period Began (millions of yr)	Main Developments of Life	Important Minerals
Cenozoic	Quaternary	1.2	Rise of civilized human beings; rise of primitive human beings and many species of mammals	Soil, peat, clay, nitrates, lignite
	Tertiary	55	Primitive forms of mammals, birds, plants; extinction of dinosaurs	Coal in western United States, Europe, Asia; oil and natural gas throughout the world; oil shale in western United States; phosphates, gold, silver
Mesozoic	Cretaceous	120	Dinosaurs and giant marine reptiles; rise of flowering trees and precursors of modern plants	Coal, oil, gas; copper in Western United States, Mexico, and Chile; lead, zinc, tungsten, molybdenum, vanadium, uranium, radium
	Jurassic	150	Rise of flying lizards and primitive birds; rise of palmlike trees	Lithographic stone; salt; gypsum; coal in Asia and perhaps Alaska; oil and gas
	Triassic	180	Rise of giant lizards (dinosaurs) and primitive mammals	Some coal, small oil and gas deposits

Paleozoic	Permian	205	Large amphibians; primitive conifers; insects in modern form	Coal, salt, potash, oil, gas; gypsum; Rocky Mountain phosphate regions
	Pennsylvanian	240	Rise of reptiles; great spread of fernlike trees, plants, insects	Coal in eastern United States, Europe, Asia, Australia; oil, gas, salt, potash
	Mississippian	290	Rise of sharks; expansion of land flora	Limestone, coal, oil, gas; oil shale in eastern United States
	Devonian	340	Rise of amphibians and fernlike plants and trees	Oil, gas, glass sand
	Silurian	380	Rise of lung fishes and air-breathing insects	Iron ore in southeastern United States
	Ordovician	470	Development of fish, land plants, corals	Oil, gas, lead, zinc
	Cambrian	540	Algae and seaweeds abundant; large numbers of crustaceans and other marine invertebrates	Phosphates; marble
Proterozoic	Algonkian	740	Primitive aquatic plants and invertebrates	Iron ore in Great Lakes region; copper in Michigan; marble, nickel, cobalt, gold
	Archean	1,740	Primordial forms of life originate	

Source: Reference 1.

Africa, Australia, Siberia, India, and China. The Permian period ended when massive land movement and widespread glacial formations resulted in the extinction of as much as 90% of the existing plant life. A new cycle of life forms developed during the subsequent period, the Triassic, when climatic conditions became more arid. Slower-growing and smaller conifers and ginko trees were characteristic of this period, and these types of plants replaced the lush, giant ferns that had flourished previously. Thus, peat formation occurred more slowly, and coal from this period is much less abundant. Virginia and North Carolina contain coal associated with the Triassic period (2). Widespread deposition of peat occurred for the second time in history beginning in the Jurassic period and reached a peak in the late Cretaceous or early Tertiary period. The appearance of angiosperms, or flowering plants, is associated with the Cretaceous period. This brought about the greatest diversity of plants ever known. In the United States the primary coal deposits formed during the Cretaceous period are found in the west. The formation of lignite and the deposition of plant debris continues in the current time period.

The cyclic pattern of coal formation in time and the deposition of coal over a large area illustrate the fact that both geography and climate are instrumental in the formation of coal. At the same time, the actual rank of coal now found in any location cannot be estimated simply in the basis of the age of the original deposit. Table 2.2 shows very clearly that there is little correlation between the geologic age associated with coal deposition and its rank today. Geologic studies show that coal is found in seams alternating with other rock types in sedimentary basins and that a coal bed itself reflects the accumulation of vegetation that grew in a warm, wet environment. Complete decay of the vegetation did not occur; rather, a less severe alteration of the organic compost resulted in the extraordinary transformation of the vegetation into coal. The damp, mild climate believed to have prevailed during both great coal-producing periods promoted a cycle of growth and decay of plants with which the rate of decay did not keep pace. The accumulation of debris in swampy areas form what we now call peat bogs. Paleobotanical studies have identified modern environments in locations across North America that typify the kind of environments that may have prevailed ages ago. Studies such as these are an integral part of a science that has the present as the only available key to reconstructing the past and to predicting and planning for what may occur again in the future. F. Rich, a geologist at the North Dakota School of Mines, has described 10 modern depositional environments that he believes serve as models of the past (4):

1. *Kettle holes:* deglaciated areas located principally in the upper regions of the Dakotas, Wisconsin, and Maine
2. *Karst lakes:* limestone sinkholes in areas of artesian water, scattered in places such as South Dakota, Missouri, Iowa, and north-central Florida
3. *Beach ridges and swale swamps:* sand ridges near shorelines; not common

TABLE 2.2 Correlation of Geologic Age and Worldwide Deposition of Coal

Period	North America	South America	Europe	Asia	Africa	Australia
Quaternary	l	l	l	l	l	l
Tertiary	a, lvb, SB, L	B, SB, L	B, L	SB, B, L	L	L
Cretaceous	a, LVB, SB, L	SB, B	B, L	b	SB	l
Jurassic	B	—	sb, b, l	a, LVB, b, sb, L	—	b
Triassic	b	—	a, B, sb, l	—	—	—
Permian	b	b	A, B	B	B, c	B
Pennsylvanian	A, LVB, B, c	b	A, LVB, B, sb, l, c	A, LVB, B	A, B, l	LVB, b
Mississippian	a, lvb, B	—	B, sb, l	A, LVB, b	—	—
Devonian	—	—	b	—	—	—

Source: Reference 3; reprinted with permission from E.S. Moore, *Coal: Its Properties, Analysis, Classification, Geology, Extraction, Use and Distribution,* 2d ed., Wiley, New York, 1940, p. 350.

Key: l, lignite; sb, subbituminous; b, high- and medium-volatile bituminous; lvb, low-volatile bituminous; a, anthracite; c, cannel; capital letters indicate major deposits, and lowercase letters indicate smaller, less-significant deposits.

4. *Impoundments:* inland depressions, caused by tectonic movements such as earthquakes or volcanoes, that block natural drainage channels, for example, Reelfoot Lake, in northwest Tennessee, and areas near the Black Hills
5. *Inland river swamps:* freshwater areas at or near sea level; typical of regions of the Mississippi River above the delta and the Powder River Basin in Wyoming; usually extensive
6. *Oxbow swamps:* along rivers where meanders abandon a prior path and leave a low, very wet region; frequent but small; for example, Horseshoe Lake, in Illinois
7. *Glacial lake basins:* abandoned glacial basins that are flat rather than deeper kettles, for example, Glacial Lake, in Wisconsin
8. *Back barrier:* barrier islands similar to beach ridges, such as the Carolina coastal plain and the Okeefenokee swamp; rare but significant peat environment because they are large and deep and could be buried rapidly if the sea rose.
9. *Deltas:* extensive marsh regions subject to marine flooding; variable size but common; "most extensive and impressive swamps of all times"; for example, Mississippi and Amazon deltas
10. *Muskegs:* bog and marsh regions, very damp and frequently poor in nutrients; extensive acreage in Canada and Ireland, lesser amounts in Alaska

The character of each of these microenvironments is not unique. Also, some, such as kettle holes and swale swamps, are quite rare; their major significance is that they illustrate the wide variety of potential coal-forming environments similar to what may have existed in past ages. However, several of these environments are extensive and provide excellent sites for geological and botanical studies. The more well-known areas include the Dismal Swamp, in Virginia and North Carolina; the cypress swamps, in Florida; the Great Lake muskegs; and the Sumatra swamp, in the East Indies (5). Besides identifying environmental archetypes, paleobotanical studies have determined that as many as 3,000 species of plants grew in the Carboniferous period (6). These plants were basically fernlike trees. Although peat is now found in many small lakes and bogs in cool, wet climates and even in more frigid regions, these isolated areas seem to be too small to ever be formed into a coal seam of significant size. Immense swamp areas such as the Dismal Swamp, which originally covered 2,200 mi^2, according to several estimates, is believed to be more typical of the size of past depositional sites (7).

The more critical difference between the present and the past eras with regard to the growth and preservation of vegetation is not the size of the region but the fact that most land masses on earth now are frigid or are subject to relatively variable climates. Geologists generally agree that centuries ago the climate was quite humid and was uniform over almost the entire earth. Furthermore, seasonal fluctuations did not occur, nor was the temperature unusually hot; it is believed that it was subtropical or tropical. Part of the evidence that supports theories about the climate in the past is the abundant vegetation documented by fossil studies. Such studies indicate that there must have been an uninterrupted period of growth, and there is no fossil evidence of seasonal rings. The similarity in the plant remains in coal taken from sites around the world also supports the notion of a uniform climate. A warm, wet climate would result in more growth and decay per unit of time. The rate of decay would be enhanced by a high temperature, but the apparent presence of large amounts of water diminished the extent of decay (8).

Two other environmental parameters may now be significantly different from conditions in the past. There is some evidence that a substantial change in the chemistry of the earth's atmosphere may have occurred as the earth developed (9). It is suggested that such a change could have had a profound effect on plant growth and cannot now be duplicated. The other factor is that land masses now are much more subject to continuous erosion than in the past. Studies show that tides, river currents, and groundwater now wash large amounts of sedimentary material away from the site of its initial deposition and spread it out over wide areas. Although transgression and regression of both fresh and marine water certainly did occur in the past, only extended periods of time between such events can account for the huge localized deposits of material that became coal.

In spite of these differences, and even if not all of the contemporary models actually progress to form coal, they nevertheless are the best illustrations of potential coal-forming environments available for study. To coal petrologists, botanists, and geologists alike, these regions are essential links to the past.

2.3 THE FORMATION OF PEAT: BIOGENESIS

Peat is one of the primary domestic fuels in Ireland and Scotland, and it is a source of natural gas. Statistics show that the Soviet Union and Canada contain three-fourths of the world's peat, estimated at 170 and 150 million acres, respectively. This acreage is equivalent to about 17 billion m^3 of natural gas (10). In the United States, peat is used commercially, but for nonfuel purposes. It has been suggested that as much as 85% is used as a soil conditioner known as peat moss or sphagnum (11).

Peat is the progenitor of coal. It consists of loosely compacted, partially decayed organic material, both plant and animal remains. Peat always contains cellulose. The color of peat varies from yellow-brown to black. It is fibrous and as much as 90% water. When it is dry (25% or less moisture), it burns with a smoky flame (12,13). Various types of peat are found around the world. Salinity and acidity of the water associated with peat deposits, the type of rock substrate, and the depth of the water during the period of deposition all affect the formation of peat. W. Francis divided peat into four categories according to its physical appearance: fibrous; pseudofibrous; amorphous; and intermediate, or mixed (14). A more common classification is based on the composition and acidity of the region and the peat itself. G.K. Fraser first used these features to identify four types of peat: highmoor; lowmoor; forest; and sedimentary, or lake (14). A recent revision of this classification scheme identifies only three types: highmoor, lowmoor, and transitional (15).

Highmoor peat develops at elevations where it is cool and somewhat humid. Because rainwater is the principal source of water in these locations, the soil is poor in nutrients (oligotrophic), particularly calcium and potassium ions, and the substrate is usually clay. The lack of mineral nutrients results in peat with an ash content of only 1–3% (15). Highmoor regions are usually moderately acidic; typically the pH is 5–5.5, although values as low as 3 have been measured. Sphagnum and grass are the principal plant species found in these areas.

Lowmoor peat consists of more diverse and more highly developed bushes and shrubs. V. Bouška includes willow and elder trees in this type also (15). As the name suggests, lowmoor peat is established in low-lying regions exposed to both surface water and groundwater. Ponds, lakes, and spring areas are examples of such regions. The substrate is generally rich in nutrient minerals (eutrophic) constantly being supplied by the groundwater and by algae, fungi, and plankton, which thrive in this environment and subsequently become incorporated into the substrate. The media is generally neutral or slightly alkaline. The abundance of mineral nutrients results in a rather high mineral content, greater than 3% and sometimes as high as 15% (15). Bouška identifies transitional peat as that which shows evidence that highmoor species developed on a lowmoor substrate. This is similar to the sedimentary peat category designated by Fraser. The other features—nutrients, acidity, and ash content—are intermediate between those of lowmoor peat and those of highmoor peat (1).

Not only is it necessary that low-lying or otherwise wet land be available for peat formation, but drainage must also be limited to minimize erosion and to maintain primarily stagnant conditions. Moving water, as opposed to stagnant conditions, results in erosion and provides sufficient oxygen to support a vigorous population of bacteria, algae, and fungi, which feed liberally on plant matter. Under such conditions, decomposition would be virtually complete, and much of the plant and animal debris would be washed away. A sufficient supply of oxygen alone can also react with organic matter in a process called autoxidation. This process would destroy much of the material before it could accumulate. As it was, the shallow stagnant water permitted only limited microbial activity, frequent covering of the surface layer with new deposits, and the precipitation of inorganic salts.

The decay processes began with the action of aerobic agents. The aerobic zone includes the first meter of depth from the surface. At the top of the zone, rotting occurred. This resulted in almost complete destruction of the organic material, releasing only CO_2 and water. Resins and waxy materials may have remained. Less severe oxidation just below the surface, called moldering, resulted in less complete and slower destruction of cellulose and the formation of complex organic compounds called humic acid or humus (16). Cellulose, which is composed of starches, is much more susceptible to aerobic decomposition than are waxes and resins. This partially decayed material piled up, and gradually subsidence of the mass occurred. As the oxygen supply and the nutrients in the water became depleted, the two aerobic processes ceased. Only the most resistant material remained at the end of the moldering stage. It consisted of carbon-rich bark, woody tissue, waxes, resins, seed coats, and spores.

An important characteristic of humus is its ability to bond with metal salts. As humus was formed, a variety of metals became concentrated in the accumulating mass. In some cases these metals were toxic to the aerobic bacteria, contributing to the cessation of the initial oxidative processes. The variety of elements in the humic material became incorporated into coal. Most of these are alkali and transition metals, such as Na^+, Mg^{2+}, Ca^{2+}, K^+, Mn^{2+}, and Fe^{2+}, but other important ions are As^{3+}, Cl^-, SO_4^{2-}, $H_2BO_3^-$, CO_3^{2-}, and NO_3^- (17).

The third biochemical stage of peat formation is anaerobic fermentation, or putrefaction. Anaerobic processes are associated with 2–10 m of accumulated debris. In the absence of oxygen, anaerobic organisms attacked carbohydrates and proteins to obtain oxygen. The hosts were primarily spores, pollen, and aquatic organisms such as algae, plankton, and small crustaceans (17). Decomposing proteins caused the accumulation of nitrogeneous products, mostly ammonia, and the pH of the media became quite alkaline. Other gases were produced as well, including methane (CH_4), nitrogen and nitrous oxide (N_2O), H_2S, CS_2, and CO_2 (17). Gels and slimes, called sapropel, resulted from these reactions. This fluid material filled voids in the remaining semisolid mass. Compared to humic peat, sapropel is rich in carbon, hydrogen, and nitrogen.

As the aqueous media became alkaline, additional metal ions precipitated and remained in the solid mass. Some ions incorporated at this time include Zn^{2+}, Cu^+, Ag^+, V^{2+}, Mo^{2+}, Ni^{2+}, and Sn^{2+} (17). The formation of pyrite is also associated with the period of putrefaction. The strongly reducing environment (absence of oxygen and the alkaline media) permitted the bacterial reduction of Fe^{3+} to Fe^{2+}. Although other ions are also reduced to lower oxidation states under these conditions, iron is particularly important. As H_2S formed and percolated through the bed, it reacted with the iron ions and formed iron(II) sulfide. Subsequent burial of the peat is believed to have caused the recrystallization of the sulfide, FeS or Fe_3S_4, to FeS_2 (17). The two forms of FeS_2 are pyrite and marcasite.

Sulfur can also remain in coal bound to organic molecules. Organosulfur compounds are not as abundant as pyrite, but they are distributed rather evenly throughout coal. Research has been under way for a number of years to more fully determine the exact nature of organosulfur species and the role they play in coal utilization. Organically bound ester sulfate groups (—C—O—S—) have been found to be the most geochemically stable form of sulfur in coal. Estimates are that they constitute 25% of the total sulfur in modern peat deposits. Evidence from related studies shows that chemically reacted portions of lignin are responsible for the production of ester sulfate in coal (18).

2.4 VARIABLE FEATURES OF PEAT FORMATION

As cycles of growth, moldering, and putrefaction alternated with subsidence and new growth, layers of humic peat and sapropel formed. Fluctuations in the water level were critical in determining the extent of each phase. After the coalification process was complete, these layers of peat had been converted into layers of more or less woody coal. Research has shown that both coastal lowlands and inland basins existed in different regions of the United States (2). As the earth's crust shifted, some lowlands became deeper and more submerged under fresh or salt water; alternatively, mud, sand, and gravel were carried into lowlands, burying a peat bed altogether. In such a situation, the coal deposit consists of a series of partings, layered seams separated by mud, clay, or sandstone.

For geologists, the elements of each cycle provide well-preserved and invaluable records of the past. A sedimentary cycle is exhibited best in formations called cyclothems. Ideally, there are several repeating layers of shale, sand, silt, and clay (19). Coal is almost always found on a base of clay. The entire sequence illustrates periods of transgression and regression of water in a depositional region. Preservation of an entire cyclothem requires that little or no tectonic movement occurred in the region to overturn it or to destroy the sequence in some other fashion. However, the complex interactions involved when the land and sea did shift are also very important.

The foregoing description of the events involved in the formation of peat may appear to be a set of long-known facts. However, at least four aspects of the discussion have been the subjects of uncertainty and lengthy debate. The controversy concerns (1) whether vegetal matter accumulated where it grew or was carried elsewhere in fragments by water; (2) whether components of a marine environment contributed substantially to peat; (3) whether lignin or cellulose was the principal plant material in peat; and (4) whether extensive compaction or a colloidal phenomenon was involved in the final stages of peat formation.

2.4.1 Sources of Plant Material

Two theories developed simultaneously during the nineteenth century and fueled the debate among one group of botanists, geologists, and chemists who maintained that coal formed from plant debris that had grown or fallen into an immediately adjacent bog or marsh area and another group who proposed that coal was formed after plant material was transported by water and deposited elsewhere, in a sea or lake, where it settled and accumulated. Accounts of the controversy written separately by E. S. Moore, T. A. Hendricks, and others make it clear that both positions were based on specific, valid observations and that eminent scientists were associated with both positions (20,21). The amount of mineral matter and other sediment generally found in coal, the presence of entire tree trunks preserved in the upright position or the presence of roots in the underclay, the size of most coal deposits, and the overall manner in which material is distributed throughout a coal seam are some of the findings that support the in situ deposition of plant material. On the other hand, the presence of marine fossils, especially those of primitive fish; the frequent absence of tree roots; the presence of upside-down trees; and the presence of the distinct series of sedimentary strata (cyclothems) that have evidence of marine transgressions are some findings that support the theory that extensive erosion was involved in carrying vegetal matter to distant sites where final deposition occurred (20,22). This is called the drift theory. C. W. von Gumbel, a German chemist, coined the terms *autochthonous* for the in situ theory and *allochthonous* for the drift theory. In 1883, nearly 100 years after the controversy first began, von Gumbel proposed that most of the geologic evidence favored the autochthonous view (20). As recently as 1917, D. White and R. Thiessen also studied the issue and likewise concluded that the autochthonous theory best explains the character of major coal deposits. These two men have made outstanding contributions to coal science, but their work has not gone unchallenged. One of their contemporaries, E. C. Jeffery, a paleobotanist, carried out extensive microscopic analyses of his own and supported the drift theory (23). In 1961, W. Francis published what has become one of the classic volumes written about coal. At that time he wrote that most geologists believed that flooding of low-lying areas did undoubtedly admit drift material, mostly mineral matter, into deposits otherwise primarily of autochthonous origin. Francis also proposed that an extreme of both processes was also possible. In what he called a mega-allochthonous situation, violent sea eruptions transported and buried material so suddenly that no aerobic decay occurred. The peat formed in this manner would have undergone only slight decay and remained highly structured. At the other extreme—mega-autochthonous conditions—aerobic processes were so extensive that no humic peat, but only finely grained and slimy sapropelic peat, was formed (24). Thus, portions of deposits or complete deposits may indeed be composed of drifted material. This dual origin and the overlapping processes accounts for the differences observed in nearly all coal seams and is reflected both in the organic composition and the proportions and types of inorganic constituents. It is believed that most coal in North America is of in situ origin.

2.4.2 Freshwater and Marine Environments

The question with regard to a marine versus freshwater environment also dealt with two aspects: whether an exclusively marine environment could be a medium in which peat formation could occur, and whether marine plants and animals contributed significantly to the development of peat being formed in a primarily freshwater environment. This topic seems to have evoked a much less vigorous debate than does the in situ versus drift issue. Moore pointed out that there was ample evidence that swamps and forests had been buried under salt or brackish water. These are called mangrove swamps, and several large areas of such swamps are known around the world. However, it was his belief that tidal currents brought such large amounts of sand into these swamps as to reduce the likelihood of peat development (25). O. Stutzer, an eminent German geologist, also addressed this question. In his work, which was translated and revised by A. Noé, it was pointed out that some coal actually originated on an ocean floor. One example is a Devonian coal seam in Neunkerchen, in Germany (26). However, several pieces of evidence militate against a purely algal theory of coal formation. First, even large masses of seaweed do not sink before they are completely decomposed. Second, the amount of fossilized marine material that has been found in coal is much too small to have come from a primarily marine environment. Third, the discovery of unique concretions called coal balls in some coal seams also supports the argument against a nonmarine origin of coal. Coal balls are hard, spherical masses of mineral matter—composed primarily of calcite ($CaCO_3$), dolomite [$CaMg(CO_3)_2$], or pyrite (FeS_2)—that contain extremely well-preserved plant fossils visible in extensive detail. Although coal balls are found in areas that are always covered with a layer of marine-derived sediment, no marine remains have been identified within a coal ball (27). Coal balls form when a mass of peat is flooded with water containing a high concentration of salt. Compaction of the mass petrifies the organic material. This is the same process that has preserved whole trees in their upright position in some locations. Although coal balls are evidence that a marine intrusion occurred, they do not constitute sufficient evidence that the original plant material grew entirely in a marine environment.

The effect of plant decay under brackish, marine conditions should not be minimized. Algae, plankton, and fish are particularly rich sources of proteins and generally contain more sulfur and nitrogen compounds than do freshwater species. These two elements are particularly significant for correlating seams with the depositional environment and for utilization of mined coal. The quantity of nitrogen and sulfur found in some peat, humus, and coal are sufficiently high that marine sources, rather than exclusively land plants and animals, must account for it. Marine grass peat in Connecticut and mangrove peat from the Whitewater Bay lagoon in Florida are two examples from depositional areas highly influenced by marine species (28).

Another important aspect of the presence of seawater during peat formation is that it is more alkaline (pH 8.0–8.5) than is fresh water. In fresh water the products of bacterial action are acidic, and fresh water associated with peat formation tends to have pH values near 7 or slightly lower. This moderately acidic condition promotes oxidation rather than reduction reactions.

Under the alkaline conditions that characterize marine environments, anaerobic processes and electrochemical reduction reactions are favored. The metabolic activity of anaerobic bacteria that results in the reduction of metal ions was described earlier. The result of the oxidation-reduction mechanisms is displayed in the elemental composition of the organic debris. The oxygen content of humic material produced in the first instance is high relative to its concentration in humus produced by anaerobic reactions, while the hydrogen content is higher in the latter case. This difference becomes very important because of its relationship to the swelling and coking properties of coal (29,30).

2.4.3 The Role of Cellulose and Lignin

The duration and scope of the cellulose-lignin controversy, which took place early in this century, were much the same as those of the in situ versus drift debate in the previous century. Cellulose is the generic name given to a class of carbohydrates produced by plants during photosynthesis. The general formula is $[C_6H_{10}O_5]_n$. The *n* indicates that carbohydrates are large, complex molecules formed from simple sugars; $C_6H_{10}O_6$ is the formula of a six-carbon sugar called a hexose. In plants, the original cell walls are composed of cellulose. Normal cellulose dissolves readily in strong acid and is degraded to glucose, one of the hexose compounds. Cellulose can be considered a polymer. A low estimate of its molecular weight is between 20,000 and 40,000; a high value is 150,000 (31,32). As plant growth occurs, some of the cellulose becomes altered to form more rigid fibers to support the plant and to enhance its resistance to microbial and chemical attack. This rigid, fibrous material is called hemicellulose. The fibers are bound together with another complex material called lignin. Lignin is actually not homogeneous material like cellulose or hemicellulose. Rather, it is composed of a variable number of different compounds that contain at least one methoxy-substituted aromatic ring. It is difficult to isolate lignin, and its characterization requires tedious, time-consuming procedures. Many compounds have been derived from lignin, including pectin and vanillin, and a number of alcohols and organic acids. Because lignin is more resistant to chemical or bacterial attack than is cellulose, two German scientists, F. Fisher and H. Schrader, proposed that lignin, and not cellulose, was the more likely source of humic acids. They believed that cellulose was readily converted to CO_2 and water very early during biogenesis and therefore had no role in the coalification process (31). Bouška cites Rakovskii as a major antagonist of the purely lignin theory (33). Francis' text gives a detailed description of the various chemical reactions related to this problem; I. A. Breger, O. Stutzer, T. A. Hendricks and others described the historical developments and summarized the arguments on both sides of the issue (34). As with the previously described controversies, only extensive research efforts and the accumulation of supporting data have made it possible to conclude that cellulosic material is much more vulnerable to oxidative decomposition and that lignin residues constitute most humic material that forms peat. All plant material, however, including protein, and the remains of microorganisms as well, is present in peat and in coal, depending on the conditions under which a given peat was deposited. The total process should be considered an integrated progression. Certainly the initial reactions involve cellulose. Some of the products of these decomposition reactions do accumulate and form part of the heterogeneous character of humus along with the more resistant lignin material (35).

2.4.4 The Colloidal Character of Peat

Until about 1917, it was generally accepted that coal formed when decayed plant debris was compressed firmly to form peat and then compacted even more severely to yield coal. This may seem reasonable because it has been documented that peat shrinks as much as 90% when buried (36). However, when this idea is translated into numbers and tested against other pieces of geologic and botanical information, it becomes more uncertain. Based on calculations regarding the rate at which peat might have been deposited, Moore concluded that 1 ft of compressed peat would be formed in 100 years; 3 ft of compressed peat yielded 1 ft of coal. In this case, a 10-ft thick coal seam would have required 3,000 years to form (36). A more recent estimate is that 1 ft of coal requires at least 6 ft of peat (37). Obviously, this means that a longer period of time was required to establish a moderately thick coal seam. Coal seams usually stretch over relatively long distances, and usually an entire seam contains the same rank of coal. If a very lengthy period of compression was required to form a seam, it can be questioned whether completely uniform conditions could be expected to have persisted for so long over such a distance (38). Evidence for dramatic shrinkage in fossilized tree and plant stems has been found, but R. A. Schmidt maintained that other evidence showed clearly that the presence of undisturbed trees, some of which continued to grow even during peat accumulation, and the absence of distortions in the strata above coal seams was not consistent with a protracted period of severe compression and shrinkage. Schmidt also argued that the process by which entire plants of all sizes have been fully preserved is consistent only with the hypothesis that peat was deposited as a colloidal material and was not subjected to unusual compaction (38).

Colloids are two-phase systems in which the dispersed particulates are larger than molecules but not large enough to be seen with a microscope. Generally the size of these colloid particulates is between 5 and 0.2 μm (5×10^{-4} to 2×10^{-5} cm). Dispersed species of this type have a very large surface area relative to their volume and have unique chemical and electrochemical properties that can be used to distinguish them from true solutions and mixtures. Two extremely significant properties of peat (and coal) that indicate that peat can be considered a colloidal material are its capacity to absorb and retain water, and its acid-base properties (39). Peat, like other organic colloids, swells and absorbs water into its microporous structure. Peat composed primarily of waxy or resinous matter swells somewhat less than does that containing more cellulose derivatives. In peat, the colloidal system is called a hydrosol. Swelling and inundation of the pore structure of decaying tissue is always accompanied by adsorption of a variety of ions. Initially, the colloid is highly protonated due to the humic acid composition. However, as mineral salts percolate through the system, the hydrogen ions are exchanged for metal or hydroxide ions. Thus, as peat formation progressed, plant remains were enveloped and preserved by this humic acid gel rich in exchangeable protons and other ions (39). Continued aging squeezed out residual water, the carboxylic acid and alcohol functional groups were destroyed, and the peat became more solid and compact. The humic gels agglomerated and penetrated into remaining woody plant tissue (39).

Some of the characteristics of lignite are that it, too, has a relatively high moisture content, will swell when soaked in a variety of solvents, and will crumble (a process called slacking) when dry. Some higher rank coals also exhibit a marked tendency to swell. The liquid is not merely trapped in the coal pores; drying a swelled coal does not completely restore its original structure. This type of hysteresis with alternate wetting and drying is indicative of a colloidal material (39).

2.4.5 Summary of Peat-Forming Processes

It is important to remember that many eminent scientists devoted lengthy portions of their careers to the effort to resolve the issues described in Section 2.4 and to come to a profound grasp of what coal is. Some of these topics are still subject to modification as new findings come to light. It is beyond the scope of this work to provide more detail, and only a complete reading of the extensive relevant literature will do justice to these topics (40).

The complex, multifaceted process described in this section constitutes what is called the biochemical stage of coal formation. Geologist D. White proposed this as the formal name in 1908, although Renault and Potonie had used this same term for some time before (41). Other synonymous terms are *biogenesis, diagenesis,* and *dynamochemical stage.* It seems apparent now that the whole spectrum of conditions from fully aerated, acidic, fresh water, to fully anaerobic, alkaline, marine water may have persisted in some places during peat formation. The primary phases of peat formation and the essential variables that impinged to a greater or lesser extent during each phase are given in Table 2.3. Movement from one portion of such a spectrum to another as conditions changed was likely to be irregular in both occurrence and dura-

TABLE 2.3 Major Interwoven Phases and the Variables that Characterize the Biogenesis of Coal

Phases	Variables
Selective destruction of carbohydrates and lignin	Water level and degree of brackishness Oxygen level Temperature
Concentration of waxes, resins, cuticles, and spores	Acidity or alkalinity of water Oxidation-reduction potential Time
Accumulation of colloidal organic and inorganic debris	Presence and type of microbes, rate of land subsidence
Burial	

tion. This irregularity had a definite effect on the type and quality, or grade, of coal that resulted from the second stage of coal formation. The formation of coal from peat is the topic of the following section.

2.5 COAL FROM PEAT: THE METAMORPHOSIS, OR DYNAMOCHEMICAL, STAGE

No sharp boundary exists to mark the point at which biogenesis stopped and coalification began. At the same time, however, peat formation did not simply pass gradually into coalification (33). Putrefaction produced a mass of woody debris and jellylike ooze in relatively large quantities of water. At some point, the depth of land subsidence, coupled with the weight and thickness of the overlying sediments, was sufficient to completely smother all forms of bacteria and fungi. This marked the end of biological, or biochemical, processes. In Bouška's view, no biochemical processes can occur at depths of more than 100 m, but others limit biochemical activity to only 10 m (42). No peat bed has ever been found below a coal seam; thus, all subsequent changes occurred in the buried peat and involved carbon enrichment. These are called geochemical, metamorphic, or dynamochemical changes.

The extent of coalification is identified by the rank assigned to a sample. Chapter 1 referred to the use of the fixed carbon value to define rank. Other characteristics of rank can be determined by laboratory analyses (see Chapter 5). These include (1) hydrogen, oxygen, and moisture content; (2) volatile matter content; (3) specific gravity; (4) presence of coking properties; (5) solubility, particularly in alkaline solution; (6) optical properties; and (7) reactivity toward chemical oxidation and reduction.

Like the formation of peat itself, the conversion of peat into coal was a continuous, irreversible process. Similarly, the progressive advance in rank does not have a number of discrete demarcations and cannot be observed directly. Although there are some breaks in the coalification process, in general one stage blends into the next (43). The graphs in Figure 1.2, which show the increase in rank of coal in relation to volatile matter and heat content, illustrate this gradual change quite clearly. It seems that coalification may be ongoing because methane is still being released underground. Demethanation of organic structures is one of the key aspects of coalification. Dehydration (loss of water) and decarboxylation (loss of carboxylic acid moities as CO_2) are also associated with coal formation. These processes are less easily detected than is the release of methane. It can be argued, however, that the presence of methane is being observed only because deeply buried coal has moved closer to the surface in more recent times; the observed phenomenon is the result only of the uplifting, not of actual "new" coalification. It is very likely, however, that in places where water percolates through a seam, alteration of the coal by leaching and by oxidation of organic components and mineral matter is in progress today.

As coalification progressed, the carbon content of the coal increased. The oxygen content, and ultimately the hydrogen content, decreased. It is for this reason that carbon content is one of the best and most direct estimates of coal rank. Table 2.4 gives the elemental composition on a moisture- and ash-free basis (see Chapter 5 for definitions of terms used in analysis) for some repre-

TABLE 2.4 Typical Elemental Composition of Peat and Some Representative Coals of Different Rank

Rank	Percent C	Percent H	Percent O	Percent N	Percent S
Peat	55	6	30	1	1.3[a]
Lignite	72.7	4.2	21.3	1.2	0.6
Subbituminous	77.7	5.2	15.0	1.6	0.5
High-volatile B Bituminous	80.3	5.5	11.1	1.9	1.2
High-volatile A Bituminous	84.5	5.6	7.0	1.6	1.3
Medium-volatile Bituminous	88.4	5.0	4.1	1.7	0.8
Low-volatile Bituminous	91.4	4.6	2.1	1.2	0.7
Anthracite	93.7	2.4	2.4	0.9	0.6

Source: Reprinted with permission from H. Tschamler and E. deRuiter, *Coal Science,* R.F. Gould, ed., American Chemical Society, 1966.

[a] Ash and moisture content constitute remaining weight percent.

sentative coals of different rank. It should be understood that each rank actually corresponds to a range rather than to some absolute value for the carbon content. It can be seen from the table that in the early stages of coalification there is a dramatic loss of oxygen content. This is the result of dehydration processes referred to above and to oxygen loss by decarboxylation. The hydrogen content remains almost constant as rank increases, and then falls sharply from low-volatile bituminous to anthracite. Hydrogen, of course, is lost during dehydration, but most of the reduction in hydrogen occurs by demethanation. The other two principal elements found in coal, nitrogen and sulfur, appear to be relatively independent of rank.

It was shown in Chapter 1 that the rank of coal is not determined by its geologic age alone. Quite early in this chapter it was mentioned that some lignite is located in the Eastern Province, a region where most of the coal deposits were formed during the Carboniferous period, 300 million years ago. Most of the coal in other locations of that region is bituminous. Similarly, coal in the west was deposited only 30 million years ago, during the Tertiary period. Not only lignite but also subbituminous and bituminous coal is found throughout the Rocky Mountain Province. Thus, coals of entirely different geologic ages can have similar chemical (elemental) compositions and physical properties, while coals of the same geologic age can have different elemental compositions and different properties. This seems to be undisputable evidence that the impact of the chemical and physical forces that operated during the metamorphic stage produce different ranks of coal totally independently of the chronological age of the deposit. The chemical and physical forces that affect dehydration, decarboxylation, and demethanation are temperature, pressure, and depth of burial. The most significant condition is the maximum temperature attained and the length of time the deposit is subjected to that temperature (44).

2.5.1 Brown Coal

It is primarily a European convention to identify brown coal separately from lignite. Two types of brown coal are recognized by coal petrographers. One is fibrous and has a high moisture content. It tears and cuts easily. The other is woody, hard, and tough when fresh but cracks from shrinkage and crumbles easily as it dries (45). The density of brown coal varies widely, from 0.5 to 1.3 (46). It is believed that the close similarity of brown coal and peat is due to the formation of only small amounts of colloidal material during biogenesis. If decaying plant matter did not remain saturated with water and hydrosols, extensive drying may have occurred before complete burial and metamorphosis took place. Failure of appreciable amounts of colloids to form and to impregnate the woody peat tissues would result in the soft fibrous texture of brown coal (47). Another explanation is that brown coal has been formed by a predominantly reductive process and other coal through a predominantly oxidative process (48). This explanation seems less likely because other data indicate that all peat was formed by the overlap of all the chemical and physical factors involved, including, always, periods of aerobic and anaerobic decomposition.

Brown coal can be distinguished from peat on a chemical basis. Brown coal contains no free cellulose, less than 70% moisture, and more than 60% carbon (49). Metamorphosis of peat into brown coal occurs at very shallow depths compared to that required for further increase in rank. E. Stach indicates that the peat-brown coal boundary is only 200–400 m (50).

The term *brown coal* is seldom used in the United States. According to Moore, the United States Geological Survey uses the term *subbituminous* for a compact "black lignite," and the term *lignite* for the low-grade brown-to-black coal that splits and crumbles on drying (51).

2.5.2 Effect of Overburden

Both heat and pressure are related to depth of burial. Bouška maintains that depth is the decisive factor with regard to rank (52). Two different observations about the relationship between the thickness of the overburden and coalification have been formulated. Schürmann developed an empirical rule by which he was able to relate the decrease in the water content of lignite to its depth (53). A more widely accepted law was formulated by C. Hilt in 1873. This law says that there is a direct correlation between the depth of a coal seam and its carbon content (54). Another statement of Hilt's law is that volatile matter content of coal decreases with its depth (55). These observations are based on valid data, but they should not be construed as explaining the totality of increased rank in such simple terms. For instance, Trotter studied a number of high-rank British coals and reported that their rank could not be correlated with their depth, as Hilt postulated (54). Strahan and Pollard, and Ashley also observed coal seams for which depth alone was not consistent with the degree of metamorphosis (56). Francis believed that failure to comply with Hilt's observations indicated the existence of special circumstances. One such circumstance is that the sample is likely to contain more sapropelic material than usual. This imposes a limitation on Hilt's rule to banded or at least highly banded coal. Another special case identified by Francis is one in which changes on the surface of the earth affected only parts of a deposit. He said that changes on the surface above a deeply buried seam nevertheless altered the depth of the seam, particularly because of the time frame involved in the alteration process. Wind or water erosion, or glaciation changed the amount of overlying sediment, and consequently the temperature and the pressure experienced by an underlying coal seam. Faulting, or folding of the earth, also increased the temperature to which a coal deposit was subjected. Francis made an important distinction, then, between a simple maximum temperature reached at any given depth, which is Hilt's law, and heat associated with movement of the earth regardless of depth of burial of the coal (55). Petrascheck believed that both of these facts played a critical role in the development of coal rank and that the amount of overburden alone had little impact on the extent of metamorphosis (54). Bouška reported that Hilt's law is always valid where temperature and pressure of the overburden were the primary factors involved in coalification; if time was more important, he believed that the law became invalid (52).

2.5.3 Effect of Temperature

It is generally observed that the rate of a chemical reaction increases twofold for a 10° increase in Celsius temperature. An attempt to correlate the increase in temperature with depth (higher than that of the earth's crust, which is assumed to be 10°C) to the rate and degree of metamorphism was made by Fox. Without developing his procedure in detail here, it is interesting to note that he concluded that 100°C was sufficiently hot to produce anthracite, that bituminous coal could be formed at temperatures between 40°C and 80°C, and that lignite could be formed between 10°C and 40°C. Using these temperature ranges, the time required for each change in rank was also estimated. At 30°C, for instance, a lignite containing 75% carbon could have been formed in 8.5 million years; at 150°C, half a million years would have been sufficient to form anthracite with 95% carbon content (57). The validity of these specific conclusions can be debated, but it is more generally accepted that the time available since deposition and decay of the plant matter has been long enough to carry out metamorphic changes in coal at much lower temperatures than was once thought possible based on laboratory modeling (58). Francis deduced a maximum temperature that he suggested was required to form coal at each rank (59):

Transition	Probable Maximum Temperature (°C)
Soft brown coal to consolidated lignite	200
Consolidated lignite to low-rank bituminous	200–225
Medium-rank bituminous to high-rank bituminous	250–280
High-rank bituminous to anthracite	280–350

These data are based on laboratory and field studies of several coal scientists, including M. Teichmüller and D. W. van Krevelen. Dehydration of colloidal humic acids can be achieved at 20°C under moderate pressure. Some fossil resins obtained from lignites have melting points as low as 90°C, but some others, such as amber, do not decompose below 300°C. Decarboxylation and loss of hydroxyl groups, reactions associated with the formation of bituminous coal, also do not require temperatures much above 200°C. Fossil cuticles and spore exines decompose near 250°C. Since these remains are found intact in bituminous coal, Francis believed that the maximum temperature was lower than 250°C. There is a marked interruption observed in the continuous change in coal structure with rank (43). This structural change is also accompanied by a sharp reduction in volatile matter. Francis concluded that these phenomena indicate that spore exines and cuticles have decomposed. Thus, he placed the maximum temperature for bituminous coal formation at 250–280°C. Petroleum cracking occurs at about 500°C. M. Teichmüller studied seams in the United States that showed that some geological formations containing coal were also favorable for the accumulation of oil (60). These seams consistently contained coal with less than 30% volatile matter. From this Francis placed the maximum likely temperature for anthracite 350°C (60).

2.5.4 Effect of Pressure

The effect of pressure with depth must be understood in terms of its relationship to temperature and to its effect on chemical reactions. With increased subsidence, both the heat of the earth and the pressure of the load increases. As the coal is heated, it becomes soft and plastic (54). The heat may be sufficient to break chemical bonds, and molecular rearrangements occur that result in the formation and release of oxygen gas, hydrogen gas, CO_2, H_2S, CH_4, and water (61). Stach emphasizes that pressure, meaning simply the weight of the overburden, also called static pressure, had a retarding effect on the progress of coalification beyond the earliest stages because the evolution of gas is made more difficult as pressure is increased (62,63).

Stutzer separated metamorphic processes into four categories (64). He did not propose completely separate characteristics or isolated effects by this division, but they are useful for describing the forces at work during the second stage of coalification.

The gradual and fairly regular variation in rank between one part of a seam and another caused by only moderate changes in temperature and depth of burial was called regional metamorphism by Stutzer (65,66). Stutzer recognized that Hilt's law was not absolute, and he was careful to point out that tectonic processes probably accounted for sites in which extensively metamorphosed coal is found adjacent to or even above the location of lower-rank coals. Such a discovery is a valuable aid for geologists trying to determine at what time in the past a fault or the uplifting of the earth had occurred. The differentiation of peat into a high-rank coal had to have occurred before the time of the disturbance that moved it into a region of otherwise lower rank.

The second metamorphic process Stutzer identified is dynamic metamorphism. This process is caused by increases in the lateral forces (shear forces) brought about by folding and overthrust that occur when mountain ranges are formed. In coal deposits where this has occurred, there is generally an uninterrupted and regular differentiation in rank across the region. The gradual alteration was ascribed to relatively small temperature increases

(65,66). Dynamic metamorphism may also account for the formation of some anthracite coal. The upheaval of the earth may have increased the pressure and elevated the temperature being exerted on a bituminous coal seam and may have brought about the final transformation to anthracite (63). In the United States, the Appalachian region westward from eastern Pennsylvania is one example of this phenomenon. In the eastern portion of Pennsylvania, where deposits are adjacent to extensively folded mountains, the coal is exclusively anthracite. To the west, toward Ohio, the land is less disturbed. No anthracite is found, and there is a gradual but definite decrease in the heat content (increase in volatile matter) of the bituminous coal as the land becomes much flatter (67). This region of Pennsylvania also illustrates the effects of erosion and glaciation. It is believed that as much as 95% of the original anthracite in the region was lost as glaciers receded. The extravagant denudation that occurred in the frigid periods following the Carboniferous period may account for an apparent interruption in the distribution of coal between the Allegheny and the Appalachian coal fields (68). Starting from a point to the south in Maryland and moving westward toward West Virginia, the same trend in the degree of coalification is observed. Anthracite is not found in Virginia, but the bituminous coal is of high rank and contains only about 18% volatile matter. In West Virginia, however, the seams contain as much as 44% volatile components. Similar patterns can also be observed in seams containing only low-rank coals, but where one portion of the region was greatly disturbed relative to another portion. Large lignite deposits and a small amount of brown coal is found in western North Dakota; moving to the southwest, eastern Wyoming has large deposits of subbituminous coal. Further west, at the foot of the Rocky Mountains, much of the coal is a mid-rank bituminous coal. The Narragansett Basin also shows progressive changes due to dynamic metamorphism (62,63).

It should not be concluded that anthracite is formed exclusively by dynamic metamorphism. Effects of the geothermal gradient alone can be sufficient to form anthracite as well as the lower-rank coals. Because coal in strongly folded regions is almost always high-rank, and frequently anthracite, there was a belief that the sole cause of final metamorphosis was tectonic folding. This view fails to recognize that the most severely folded rocks usually sank to extreme depths before the folding actually occurred, and that the coalification was achieved by the heat associated with the depth (69,70). Data collected by M. and R. Teichmüller have demonstrated how this conclusion was reached (71).

The third category of metamorphic effects includes the specific changes brought about by the heat of molten rocks. The particular importance of contact metamorphism, as this is called, is that it can alter the mineral content as well as the structure and rank of the coal. This phenomenon involves an extreme increase in temperature (a minimum change of 500°C and a maximum change of 1,000°C are not unusual) accompanying a volcanic eruption (72,73) or an extremely exothermic oxidation reaction involving mineral matter (70). The percent of carbon in coal is increased by the heat from the intrusion, but unlike the processes involved in regional or dynamic metamorphosis, these changes occur rapidly, and the effects are limited to a region close to the site of the intrusion (74). The limited range is explained by the fact that the magma flow cools and hardens quickly, and itself is a poor conductor of heat. Because no sign of a tarry residue remains in areas exposed to igneous intrusions, there is question as to what happened to the volatile components of the coal. The best explanation is that the sudden surge of heat rapidly superheated this material and distilled it away from the coal. Francis described contact metamorphism as analogous to destructive distillation in a laboratory. In a laboratory, heating coal to 150°C is enough to begin to destroy its microstructure, and at temperatures between 300°C and 400°C, "serious disorganization" of the matrix occurs (73,74). In fact, natural coke formation is usually the result of igneous intrusion. Beneficial improvement in rank (roasting) can occur if the intrusion occurred near a seam rather than moving into it. This effect has occurred in several areas of the west. The Wolf Creek coal field, in Colorado, is one example (75). Other effects include increased compaction and density, decreased moisture content, and increased coloration. An increased number of cleats frequently resulted and the columns of coal that separated from one another are called finger coal (76). Because most beneficial effects of contact metamorphism are rare and are only localized, this process has little importance as a means of useful carbonization of low-rank coal. Describing the effects as localized does not mean that magma flow was necessarily limited in range, but only that the effect did not extend far away from the edges of the flow. Moore pointed out that in the Cerrillos coal field, in New Mexico, a 400-ft long magma sill produced a low-grade anthracite along its entire length, but only a few inches into the body of the seam (77). On the other hand, contact with molten rock that invaded a coal bed at frequent but irregular intervals could disrupt the quality of the deposit to such an extent that the entire seam was rendered worthless. The most extreme case of igneous metamorphism is the formation of graphite. Graphite is commonly found in an area immediately adjacent to a coal seam, and deposits of the two materials may even surround one another. The lower portion of the Narragansett Basin was metamorphosed largely to graphite by contact with the igneous intrusion that caused widespread formation of granite from feldspar and quartz, while anthracite coal and volcanic rocks were formed in the more northern section of the basin (78).

The fourth category outlined by Stutzer is chemical metamorphism. Chemical changes, particularly those associated with mineral oxidation, frequently produce quantities of heat. Oxidation of pyrite to iron oxide or to sulfuric acid may even cause spontaneous combustion of the coal. As a means of altering the rank of coal on a large scale, this type of metamorphism is of little value. Of course, the character and quality of a coal seam or the mined coal may be reduced if charring occurs as a result of oxidative processes. This is a serious problem for coal producers and coal consumers alike, who must ship or store large stockpiles of coal and whose work requires that the quality of the coal remain constant.

2.5.5 Metamorphism and the Structure of Coal

For more than a century scientists have sought to define an exact structure for coal. A reasonable approach to this problem was to determine the structure of cellulose, lignin, and peat. Changes in the composition of each rank of coal can be determined and models of the structures at each stage can be developed from the composition. R. A. Mott extended most of the earliest investigations regarding the chemistry of coal formation. He demonstrated that varying amounts of carbon dioxide, water, and methane constitute the volatile products during the conversion of peat through the various ranks of coal. As the rank increases, however, the amount of methane becomes increasingly large. According to his data, the formation of semianthracite (92% C) from a high-volatile bituminous coal (86% C) produces six times more methane than carbon dioxide and four times more methane than water (on a molar basis). Final anthracite formation (95% C) yields no CO_2 but does produce 36 moles of CH_4 per mole of water (79). Today more than ever before, chemists are eager to develop the most comprehensive description possible for the structure of coal. There are some discrepancies in Mott's data, but both he and Van Krevelen have made outstanding contributions to all subsequent efforts with regard to the relationship between rank and structure (80). Based on Mott's analyses, Francis concluded that anthracite can only be formed by methane production. This process requires that the organic ring structures condense into highly aromatic units. Van Krevelen carried out analyses similar to Mott's from which he also concluded that coalification is marked chemically by the increasing degree of aromaticity and polymerization due to dehydration and loss of methane (81). The importance of knowing at least the basic structure of coal in order to carry out successful gasification or liquefaction reactions cannot be stressed too much. Determining how oxygen, nitrogen, and sulfur heteroatoms are bound in coal has required a considerable amount of research effort. Similarly, the determination of the number of aromatic units per coal molecule and the average molecular weight of these units is another area of critical research. These aspects of coal chemistry are addressed more fully in Chapter 3.

2.5.6 Metamorphism of Coal and the Occurrence of Petroleum

The information in Table 2.1 shows that petroleum and natural gas have been formed in every geological period since the Ordovician except one: the Silurian. It is interesting to discuss briefly what relationship exists between the formation of petroleum and the deposition of peat and the formation of coal.

Without going into detail, it is believed that oil deposits were formed largely from marine plants and animals in a region of sedimentary formations in close proximity to seawater. After a lengthy period of moderate temperature and pressure, the organic matter was converted into liquid hydrocarbons and methane. Although this material could migrate by percolating through sedimentary material, tectonic motion also squeezed the liquid and gas into large pockets, or reservoirs, usually of sandstone or limestone. If large fissures opened, or if no impermeable rock lay over an oil formation, the fluid and gas escaped at the surface (82). In some respects this process of formation and collection of oil and gas in trapped pockets is analogous to the formation of coal seams. Some differences, as we have seen, are that coal formed in large deposits primarily from land vegetation in freshwater areas, while petroleum formed in smaller quantities in a more marine environment. These differences are not so exclusive, however, as to preclude the occurrence of petroleum from areas near coal seams or the migration of petroleum into regions where coal is located. The site of the first American oil well, drilled by Colonel E. A. Drake in 1859, is the ghost town Pit Hole, Pennsylvania, near Oil City, which is located near the northwestern limit of the Pennsylvania bituminous coal fields. Since dynamic metamorphism that may have transformed coal in eastern Pennsylvania to anthracite had little effect on the coal in northwestern Pennsylvania, neither was the oil and the natural gas in the region subjected to destructive temperature or pressures. It is abundant and of high quality. Studies carried out by Thom, White, Roger, and M. and R. Teichmüller over a 70-year period beginning in 1860 have shown that petroleum rarely exists in regions where metamorphism has advanced coal beyond 60% fixed carbon content and 37–39% volatile matter. The reason for this is that thermal changes associated with increased metamorphism of coal cause "cracking" and distillation of petroleum. The weight of the oil is decreased as methane or other light gases are cleaved and evolved. In the extreme (igneous metamorphism), bitumen is produced (83). Although oil is certainly not found near all regions of low-rank coal, in Texas large quantities of both oil and lignite are found. This may represent a unique region where geologic conditions were optimum for both types of material to be formed.

Oil shales are not related directly to petroleum-bearing rocks. Solid organic particles called kerogen, rather than liquid hydrocarbons, have been incorporated into the shale. Kerogen is derived from decayed primitive organisms, but unlike petroleum or bitumen, it is insoluble in virtually all common organic solvents.

2.6 DESCRIPTION AND PHYSICAL CLASSIFICATION OF COAL

Thus far we have seen that comparative descriptions of coal can be made on the basis of any number of its variable characteristics. The extremely heterogeneous nature of coal has both positive and negative aspects for the industrialist and the research scientist. The numerous perspectives from which coal can be studied and characterized provide a great deal of flexibility and many avenues of research. This same multiplicity of variables has unavoidably resulted in ambiguous data and terminology. Some descriptions are not as sophisticated as others, but all have been useful. In the generations preceding our own, when coal was a common home heating fuel, it was sufficient to know whether or not the coal to be purchased was "hard" or "soft." Hard coal was clean to hold. It burned long and hot, and did not leave many cinders to shovel out from below the furnace grate. It was also more expensive. Soft coal soiled one's hands, was smoky and sooty when burned, and left a large residue of cinders. It was also more affordable. Anthracite versus bituminous is a simple classification of coal by rank. It certainly was not a distinction based on analysis, but it was a distinction that was very real and very practical. Early industry required a more careful delineation of heating power. The analysis of calorific value with rank revealed more specific differences than were needed for domestic heating and led to the designation of separate groups within each rank. It is critical to bear in mind that no one classification scheme developed autonomously. The whole concept of rank grew out of the recognition by geologists that the geologic history of coals produced specific, identifiable differences. Proximate and ultimate analyses played a significant role in the formation of the concept of rank. Various uses of the information have been made according to need.

In Section 1.4.2, we distinguished the rank of coal from the type (banded and nonbanded). The same two types can occur in all ranks of coal, and within a seam of a given rank of coal, both types may be found. Thus, any seam can be described by two mutually exclusive modes of classification: rank and type. The type of coal refers specifically to the botanical origin of the plant material and tells us why the metamorphosed coal has the kind of chemical constitution and physical properties that it does. It indicates also whether one or more cycles of peat deposition occurred. Rank is indicative of the degree of metamorphism to which the plant debris was subjected, regardless of its botanical origin. Whatever the original chemistry was, determination of rank tells us what happened to the chemical composition in time under the influence of heat and pressure. It is the result of those events and the interaction of all the variables to which we have been referring throughout this chapter.

The systematic study of coal as an organic sedimentary rock is the scientific discipline called coal petrography. Coal petrography encompasses nonbanded coals as well as banded coals, but nonbanded types are less important commercially, and this discussion will not address them in as much detail. The interested reader is directed to the work of H. J. Rose (84).

In England in 1854, the famous Torbanite Case raised the question as to whether boghead coal is indeed coal. The application of microscopic analysis in the resolution of that case revolutionized coal science (85). Before then, the study of coal was limited to the examination of coal as a whole to develop theories to explain the evolution of peat, coal, and petroleum, and to identify the plant remains still visible in coal. F. Link and F. Schulz were the first to use digestion techniques routinely to render coal more transparent for microscopic analysis and to isolate particular components for microscopic analysis (86). Since then, an entire gallery of coal scientists have made the use of microscopic analysis the single most important tool to the coal chemist and the coal geologist. Stach and Van Krevelen have both described the work of those who have most strongly influenced the progress and direction of modern coal petrography (87,88).

Two systems of petrographic analysis are used most commonly to describe the banded species in coal. One was developed in the United States and the other in England. Most European countries adopted the latter system, and in its revised form it is now used by all coal petrographers. The American system requires the use of an obsolete and limited technique, and is primarily concerned with identifying the plant constituents that are independent of the rank of the coal. However, this system is still is use and is recognized by the International Committee for Coal Petrology (ICCP). Both systems are described in the sections that follow.

2.6.1 Thiessen–Bureau of Mines System

Working at the United States Bureau of Mines, two geologists, R. Thiessen and D. White, developed a system of classifying coal according to botanical principles, the nature of the original plant matter. Until microscopic methods became available, geologists were limited to describing hand samples of coal as banded or nonbanded and indicating whether the bands had "bright" or "dull" components. Using the microscope to study these bright and dull bands, Thiessen and White found that light transmitted through thin sections of coal revealed that the bright bands were more transparent and had a woodlike character, while the dull bands were less transparent and appeared as fine, formless particles. They published their work in 1913, and in 1920 Thiessen published a second volume containing a large compilation of data from his microscopic analyses of bituminous coal (89). Thiessen named the bright, woody bands anthraxylon, from the Greek words for coal, anthrax, and wood (*xylon*). This material is translucent and appears red. Thiessen deduced that anthraxylon included cellular tissues from stems and branches in a smooth, black, lustrous, homogeneous mixture.

Thiessen called the dull, gray-to-black, translucent or opaque material attritus. Attritus may compose the bulk of some coals and consists of a variety of plant remains, including fats, oil, resins, spores, and mineral matter. In thin sections, some attritus transmits light; it is called translucent attritus. Opaque bands less than 37 microns wide is designated opaque attritus. Thiessen called the third major component fusain. In order to correspond more closely with the European system, fusain incorporates large bands ($> 37\ \mu m$ wide) of opaque attritus (90). White spots are also visible throughout a specimen and indicate open cells.

All banded coal contains more than 5% anthraxylon. Bright, semisplint, and splint coal are the types of banded coal. They are distinguished according to the amount of opaque attritus: bright, less than 20%; semisplint, 20–30%; splint, more than 30% (91).

Nonbanded coal contains less than 5% anthraxylon. Cannel and boghead coal are the two nonbanded coals. Cannel coal is dull and has a greasy luster (92). The attritus in cannel coal is composed of spores, while the attritus in boghead coal is derived from algae (92). Viewed using transmitted light microscopy, spores appear quite yellow and are frequently large ellipsoidal bodies. Typically, spores will have a visible line through the length of the ellipse where the edges were pressed together. Algal components are somewhat more orange and may have a streaked appearance. It has also been shown that algal material fluoresces but that spore remains do not (93). This system of characterization and nomenclature has been adopted officially for use by the United States Bureau of Mines and is generally known as the Thiessen–Bureau of Mines System. Another agency, the United States Geological Survey (USGS), adopted a different system for describing the physical character of coal. The USGS system does not involve microscopic analysis but it is closely related to the European system, known as the Stopes-Heerlen nomenclature. The Stopes system is discussed in the next section, the USGS system in Section 2.6.3.

2.6.2 Stopes-Heerlen System of Nomenclature

In 1919, M. Stopes, a British paleobotanist, developed a system of classifying coal by analogy to the methods used for inorganic rocks. The fact that she was concerned only with bituminous coal was one of the major criticisms of this system (94). Subsequent development of the method has extended the application of its terminology to both higher- and lower-rank coals. Stopes focused her attention on the physical properties of the components of coal and the quantitative variation of these components with rank.

Geologists recognize three general types of inorganic rock: sedimentary, igneous, and metamorphic. The appearance of these types is sufficiently different that they can be distinguished with the unaided eye. By analogy, Stopes identified four "rock ingredients" in humic (banded) coal that are visible macroscopically. These ingredients are called coal lithotypes, a term used first by C. A. Seyler, also a paleobotanist. Stopes identified four lithotypes, which she called vitrain, clarain, durain, and fusain.

Vitrain is a bright and glossy band. It can range from a few millimeters to several centimeters thick. It originates from wood or bark, and is the main component of all the bright bands in coal.

Clarain is smooth and silky, interlaminated bright and dull. No specific origin is known for clarain, and Moore suggested that it is not a separate lithotype from vitrain (95).

Durain is dull, black or gray-black, and very hard. It is compact and leaves a sooty residue when handled. The distinction between clarain and durain is not sharp. In 1957, two more lithotypes were added to the four Stopes had identified. Clarodurain and duroclarain are considered intermediate lithotypes, and reflect the uncertainty and the limitation involved in using only visible appearance to identify coal components (96).

Fusain had been known for more than 100 years at the time Stopes incorporated that term into her system of nomenclature. It is a unique, charcoallike component of coal. It is porous and friable, and its cells frequently contain mineral matter. The origin of fusain has been the subject of much speculation and debate. Because it appears as though it has been charred, many people believe it was formed by volcanic heat or forest fires. It is frequently called fossil charcoal. Thiessen proposed that the action of certain thermophilic bacteria are capable of causing the development of locally high temperatures that could cause similar charring (97). Thermophilic bacteria can thrive at temperatures above 100°C, but it is not clear whether they actually produce high temperatures. The origin of fusain is complex, and the question remains unresolved (98).

On the basis of these descriptions, an approximate correlation can be made between Stopes' and Thiessen's terminology, even though the former is based exclusively on the physical appearance of coal without the use of a microscope. Both use the term *fusain* synonymously. Thiessen did not identify durain in banded coal. The material he described as opaque attritus corresponds closely to durain. Vitrain seems to be equivalent to anthraxylon, and clarain is similar to translucent attritus. Table 2.5 shows the correlations between the two systems and some specific properties of each lithotype in a typical high-volatile bituminous coal. Notice that the formal title of the European system of coal nomenclature that officially adopted these terms is the Stopes-Heerlen system. Geleen, translated as Heerlen, is the city in the Netherlands where the Congress of Carboniferous Stratigraphy met in 1953. It was at that conference that an international nomenclature was established by the International Committee for Coal Petrology (ICCP). The complete description is represented in a summary by G. Cady (99). The American system and the European system of Stopes developed independently, and naturally some confusion resulted. The ICCP compiled a glossary of all terms in use in an effort to minimize the confusing duplicity of definitions. Although the Stopes-Heerlen system is more common, American and Soviet terminology is included in the glossary. This work was begun in 1953 and was agreed upon internationally in Madrid, Spain, in 1960. Concerned that the Stopes-Heerlen system makes no provision for even the major subdivisions of rank, a group of American geologists led by W. Spackman proposed that the three maceral

TABLE 2.5 Approximate Correlation of Nomenclature in the Theissen–Bureau of Mines System with the Stopes-Heerlem System

Common Name	Coal Type in Theissen–Bureau of Mines	Lithotype in Stopes-Heerlen	Properties of Typical High-Volatile Bituminous Lithotypes
	Anthraxylon	Vitrain	Relative hardness: 2 ‡$\rho \sim 1.3$ Lowest in ash 30–35% volatile matter Uniform, shiny black bands
Bright coals (U.S. & G.B.)	Fusain	Fusain	Relative hardness: 1 $p \sim 1.35$–1.45 (soft) $p \sim 1.6$ (hard) Highest in ash 10% volatile matter Charcoallike, forms dust
Semisplint, (U.S.) Dull (G.B.)	Translucent Attritus	Clarain	Relative hardness: 3 $p \sim 1.3$ Moderately low in ash 40% volatile matter Laminated shiny and dull bands
Splint (U.S.) Dull (G.B.)	Opaque Attritus	Durain	Relative hardness: 7 $p \sim 1.25$–1.45 Moderately high in ash 50% volatile matter Dull, nonreflecting, poorly laminated

Source: Reference 98; L. Petrakis and D.W. Grandy, "Coal Analysis, Characterization and Petrography," *J. Chem. Ed., 57,* 689–694 (1980).
‡ρ = specific gravity.

groups be termed *suites*. The vitrinite suite, for example, would permit the reference to subdivisions xylinoid, vitrinoid, and anthrinoid, corresponding to the three major ranks, lignite, bituminous, and anthracite. This proposal has not been formally accepted by the ICCP and is not used widely except in the United States steel industry (100).

2.6.3 United States Geological Survey System

The United States Geological Survey (USGS) system was developed by J. Schopf, who was associated later with Ohio State University. This system is the most important macroscopic system of coal nomenclature. The USGS considers only two banded lithotypes: vitrain and fusain. Vitrain is identified as the bright-banded component; fusain is soft, sooty, and dull. All non-banded features of the coal are referred to as attrital coal. This corresponds to clarain and durain in the Stopes-Heerlen system. No microscopic analysis is used for the USGS description of coal; rather, the bands are described by their thickness and their abundance, and attritus by its luster. Each of these characteristics is subdivided in a semiquantitative fashion. The subdivisions are shown in the following scheme (101):

Thickness	Lithotype	Abundance	Attrital Character
Thin: .5–2 mm		Sparse: < 15%	Bright
Medium: 2–5 mm	Vitrain	Moderate: 15–30%	Moderately bright
Thick: 5–50 mm	Fusain	Abundant: 30–60%	Mid-lustrous
Very thick: > 50 mm		Dominant: > 60%	Moderately dull
			Dull

A sample of coal is described by selecting the appropriate term for each component, such as "thick vitrain moderately distributed in moderately bright attrital coal." This is a somewhat cumbersome system and is used primarily for a rapid description of a field sample (101).

2.6.4 Other Descriptive Features of Coal

Two other features of coal may be readily apparent in a field sample and are useful in a thorough description of the seam itself as well as the rock associations. Bands of silt or clay called partings are frequently present in coal. These bands represent mineral sediments carried into a peat deposit before complete burial occurred. The mineral matter itself has no value. The presence of wide or frequent partings diminishes the quality of the coal, since the partings cannot be avoided during mining operations and simply contribute to the ash content of the coal. Bone coal is a name given to coal that has many finely grained partings across the width of the seam.

Pyrite may also appear as a distinct band in coal. Usually pyrite bands are thin, but they also severely diminish the quality of the coal. A formal description of a coal seam includes an indication of pyrite bands.

The ASTM manual includes a standard that defines the terms used to describe macroscopic observations of coal seams in the field and in hand specimens, and the terms used in microscopic analyses as well (102).

2.7 PREPARATION OF THIN SECTIONS

When microscopy was first introduced to coal analysis, the principal techniques available to coal scientists were either to boil coal in kerosene or to digest it in a strong acid mixture to produce a transparent specimen (103). The technique of slicing thin cross sections of plants for microscopic analysis had been used by botanists for many years. In 1910 E. C. Jeffrey, an American coal petrographer, developed a similar technique for preparing coal samples (104). Jeffrey's method required two steps. First, the coal was soaked in an aqueous solution of potassium hydroxide or sodium hydroxide to soften it. This sometimes required several days and a hot solution. The use of aqua regia followed by an alkaline soak was another alternative. Jeffrey found that phenol would also soften coal and did not cause the coal to swell the way hydroxides did. After the coal was soft, the second step removed the mineral matter by dissolving it in hydrofluoric acid. Usually the softened, demineralized coal was encased in a cellulose binder, called Schering's celloidin, to hold it stiff. The coal then could be sliced with a microtome into as many sections as were desired. Machines could be used to slice the coal or to grind it down into a thin section. However, this was commonly done manually because the maximum thickness is 10–12 μm for bituminous coal and only 1 μm for anthracite (105). Thiessen and others modified the basic method outlined by Jeffrey, but in every case, it was necessary to remove the mineral matter from the sample (106). Thus, no information about the inorganic content of the section was obtained, and some organic matter was necessarily lost in the soaking step. This was especially true of softer low-rank coals.

Thiessen later combined a process of grinding coal with the digestion method. The principal advantage of this technique was that it did not require the removal of mineral matter (107). The method was improved further so that digestion was not necessary at all. The procedure involves impregnating the coal with a binder and using successively finer abrasives to achieve the desired thinness (108). This permits the preparation of thin sections for low-rank coals, but high-rank coals cannot be ground sufficiently to become transparent (109).

To study thin sections, it is necessary to have an ordinary transmitted-light microscope. If a thin section is prepared with skill, the various components are easily distinguished. Not only are the various structures clearly different, but the colors of the various species are very distinctive. This characteristic of transmitted-light microscopy is very useful, but unless photographic records of the specimen are also made in color, this advantage is lost. Because of the difficulties inherent in this technique and because of the advanced techniques available using polished sections, this method is no longer used for routine analyses even in the United States.

2.8 PREPARATION OF POLISHED SECTIONS

The use of the metallurgical technique of examining polished surfaces under reflected light was first applied to coal by H. Winter in Germany in 1919 (110). In general, polished surfaces are easier to prepare, although some petrographers believe that they reveal less information than does a skillfully prepared thin section. This method is especially useful for studying anthracite, which is too dense to be softened and cut into sections thin enough to transmit light.

If a block of coal is chiseled from a carefully marked section of a seam and is polished, it is a true polished section. Because complete characterization of the width of a seam requires a number of such blocks, this method is quite tedious and is reserved for special situations. A polished, pelletized sample can be prepared from powdered coal. The sample is mixed with a resin in a mold. It may be allowed to harden naturally, or it may be held under pressure in a vice until it hardens. After the mixture has hardened, it is polished to a mirror surface with a series of fine abrasives. The French petrographer Duparque was an advocate of "simple polishing," as this is called. Stach, his German counterpart, preferred to etch the polished surface as well, to bring out additional structural relief. The ASTM describes a standard method for preparing pelletized samples (111).

In 1925 Stach introduced the technique of examining polished blocks with reflected light using oil immersion. With advanced optics, pioneered mainly by the Leitz Company in Germany, oil immersion reveals additional detail about the components in coal and makes it possible to observe the rate of change of reflectance across the surface of a specimen (112,113).

A metallographic microscope is necessary in order to study unetched polished surfaces. A standard magnifying lens or one suitable for use with oil is also required. The measurement of reflectance requires an extremely sensitive photometric device that permits the use of polarized light. The specimen is placed at right angles to the optical axis. The structure of components is distinguished by the differences in light intensity reflected from the surface, and photographic reproduction is true to what is actually observed. Waxy material is most clearly visible, but when the sample is also etched or viewed with oil immersion, even very small differences in surface features become visible (114). A crossed hair located in the microscope ocular is superimposed on the specimen, and the components can be identified at as many as 500 intersections on the grid (114,115). This technique is called modal analysis or a point count. The instrument is calibrated against standards of known reflectance. The intensity of the beam of light reflected from the surface of the coal specimen is compared to the standards. The reflectance of 50 microscopic units per specimen and of two specimens per sample are measured (111).

2.9 THE MICROSCOPIC CHARACTER OF COAL

During the 1920s three microscopic units in coal were recognized. Stach believes that this marked the first landmark development in coal science. The second landmark was the discovery that the light reflected by components could be related directly to the rank of the coal (112). This ability to correlate microscopic character with rank has been of fundamental importance to coal science, and reflected light is used now almost exclusively.

In inorganic rocks, species that have a definite chemical composition and uniform character are called minerals. Minerals are most often crystalline. Granite, for instance, is a type of igneous rock. It is composed of unequal amounts of three minerals: quartz, feldspar, and mica. Microscopic analysis of polished sections of coal reveal that it is also composed of varying amounts of different components. Three units were identified, and the amount of each one was found to vary widely from coal to coal. In 1935, Stopes named these microscopic units macerals to correspond to minerals in other rocks. The word *maceral* comes from Latin and means to soften or weaken, with or without heat, or to wear away. The three maceral groups are called vitrinite, liptinite (or exinite), and inertinite. The analogy between minerals and macerals is not a strict one. Maceral groups are not crystalline. They were formed from different plant materials and possibly in different environments. Consequently, they contain many different types of chemical substances. Since they were altered to various extents by geochemical forces after burial, their composition and properties vary (116). It is interesting to note that as metamorphosis progressed, the apparent distinctions between the maceral groups became less clear (117). At the same time, the same lithotype in two samples may still reveal very different microstructures.

Vitrinite is derived from the wood or bark of the original vegetal matter. Vitrinite is typically a shiny and glasslike in appearance. It is the predominant maceral found in coal, and Stach relates that "petrographers have convinced coal chemists that vitrinite is the only maceral which allows rank to be determined exactly" (118). Vitrinite is the only important component of vitrain (116,117). It is particularly rich in oxygen and has a moderate amount of hydrogen and volatile material. It is the important component of coking coal. Observed in a thin section, vitrinite is red; in a polished section it reflects sufficient light to appear gray. Vitrinite has two maceral units: collinite and telinite. Collinite is amorphous and was formed when cell walls were thickened with gelatinous humus. Telinite still exhibits distinct cell structure.

The second maceral group, liptinite, was formed from the waxy resinous debris of spore exines, algae, and cuticles; hence, its maceral units are resinite, sporinite, alginite, and cutinite. Liptinite is a hydrogen-rich maceral. It consists primarily of aliphatic constituents and has a larger volatile matter content than does vitrinite. Liptinite is also reactive during carbonization. It appears dark gray in reflected light. Vitrinite and liptinite together constitute 75% of the organic material found in a typical bituminous coal.

Highly oxidized or weathered wood and bark, which is relatively unreactive toward carbonization, constitutes the third maceral group, inertinite. The lithotype fusain is composed of inertinite. Inertinite is rich in carbon; it is highly aromatic and has little hydrogen or volatile matter content. There are five maceral units in the inertinite group: macrinite, micrinite, fusinite, semifusinite, and sclerotinite. Macrinite and micrinite are protoplasm residue. Macrinite is described as massive ($> 10\ \mu$ in diameter); micrinite is granular (2–10 μ in diameter). Fusinite is chemically or biochemically oxidized woody tissue in which cell structure is visible. Semifusinite is similar but appears to be less severely oxidized. Sclerotinite is composed of the resistant remains of fungal attack on the woody plant tissue. Inertinite is white or light yellow-gray in reflected light, and usually appears to have some "relief" at the edges. Table 2.6 summarizes this classification scheme and identifies some important properties of each group.

Microlithotypes are combinations of macerals. Clarite is composed of vitrinite and lesser amounts of liptinite. Duroclarite contains small amounts of inertinite as well as vitrinite and liptinite. In the microlithotype clarodurite, however, inertinite is the primary component, along with a lesser amount of liptinite and even less vitrinite (119).

Chemical analysis of macerals is also very informative. The properties of the vitrinite and liptinite macerals and of micrinite depend on the rank of coal in which they are found. The fusinite maceral, however, appears to be independent of rank. The analysis of macerals of the same rank of coal has been carried out by Tschamler and de Ruiter (120). Their analyses and a typical fusinite analysis are summarized in Table 2.7.

2.10 APPLICATIONS OF MICROSCOPIC ANALYSIS

As we have seen, both transmitted and reflected light can be used to study the maceral components of coal. Using transmitted light, the various visible colors can be associated with particular organic species. Also, the spatial relationships among the components is preserved. The usefulness of this technique is severely limited, however, because there is no correlation between the intensity or the wavelength of light transmitted through a specimen and the rank of the sample. The preparation of thin sections is tedious and difficult. However, unless many sections are prepared, this technique does not permit conclusions that are representative of a seam.

Because pelletized specimens are prepared from a "channel sample"—coal

TABLE 2.6 Origin and Importance of Macerals and Maceral Groups

Maceral Group	Maceral	Origin	Group Properties
Vitrinite	Collinite, telinite	Wood or bark	Principal constituent of coal, readily hydrogenated, oxidized easily, relatively oxygen rich
Liptinite (exinite)	Resinite Cutinite	Plant resins Needles and leaf cuticles	Yields greatest quantity of by-products when carbonized
	Sporinite Alginite	Spores Algae	Readily hydrogenated in coals with more than 25% volatile matter, relatively hydrogen rich, tough (restricts dust formation)
Inertinite	Sclerotinite	Fungal remains	
	Micrinite, macrinite	Granular matter from protoplasm	Relatively inert, not easily hydrogenated or oxidized
	Semifusinite	Wood or bark severely oxidized before burial	An important additive in coke

Source: Reference 117; reprinted with permission from L. Petrakis and D.W. Grandy, "Coal Analysis, Characterization and Petrography," *J. Chem. Ed., 57,* 691–692 (1980).

TABLE 2.7 Analysis of Macerals of the Same Rank

	Elemental Composition				
Maceral Group	Percent C	Percent H	Percent O	Volatile Matter[a]	Aromaticity[b] (f_a)
Exinite	84.1	7.0	6.3	66.7	0.62
Vitrinite	83.9	5.5	8.0	35.2	0.77
Micrinite	85.7	3.9	8.0	22.9	0.89
Fusinite	91.5	3.2	4.3	12.8	0.94

Source: Reference 120.

[a] Percent by weight.

[b] Fraction of all carbon atoms contained in aromatic units as determined from nuclear magnetic resonance (NMR) analysis (see Chapter 3).

chipped from the whole width of a seam in a constant width and depth—they are considered representative of the entire seam. The sample collected is homogenized and ground to a small particle size so the spatial interrelationship of components is lost. Pelletized samples are easier than thin sections to prepare, and many can be prepared simultaneously. The use of reflected light also permits the use of higher magnification than is used for thin sections (600 × versus 200–300 ×).

In 1932, two coal petrologists, E. Hoffman and A. Jenkes, published data showing that coal rank could be determined rapidly and accurately by the reflectivity of the vitrinite maceral (112). In practice, monochromatic light (546 nm) is used. Macerals are distinguished by the differences in the intensity of the incident light reflected from the surface. The reflectance value R_0 is the proportion of incident light reflected by the coal surface. It is usually expressed as the percent reflectance, or reflectivity. The reflectivity of 50 vitrinite macerals per specimen and of two specimens per sample is usually measured. Because samples of pure vitrinite can be hand-picked, reflectance data can be correlated directly with the proximate and ultimate analyses of the sample. This kind of data has become the most commonly used by coal petrographers. Table 2.8 shows how vitrinite reflectance and rank can be correlated. Notice that the values in Table 2.8 are quite small: 3% or less of the light incident on the specimen is reflected. This illustrates clearly that coal, which appears black to the unaided eye, is indeed a highly light-absorbing material. The average reflectance for coal is only 1%.

A similar technique for using reflected light to determine rank and maceral composition of coal was developed by C. A. Seyler. He also observed that the maceral components of banded coal have characteristic reflectance values. He defined the term reflectance number N_r, which has integer values from 1 to 9. Pure fusinite was assigned the highest reflectance number, $N_r = 9$. An example of Seyler's data is shown in Figure 2.1. These graphs show the percent reflectance versus N_r for a sample bright clarain and for a durain sample. The

TABLE 2.8 Oil Reflectance Limits by Rank by ASTM Analysis

Rank	Maximum Reflectance (%)	Random Reflectance (%)
Subbituminous	−0.47	
High-volatile bituminous	C: 0.47–0.57 B: 0.57–0.71 A: 0.71–1.10	0.50–1.2
Medium-volatile bituminous	1.10–1.50	1.12–1.51
Low-volatile bituminous	1.50–2.05	1.51–1.92
Semianthracite	2.05–3.00 (approximately)	1.92–2.50
Anthracite	> 3.00 (approximately)	> 2.50

Source: Reference 121.

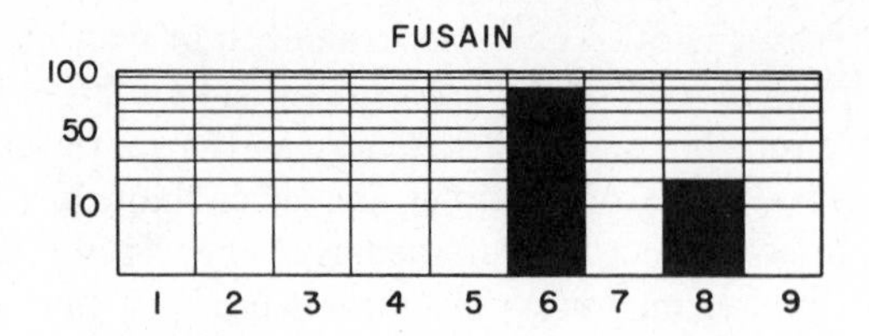

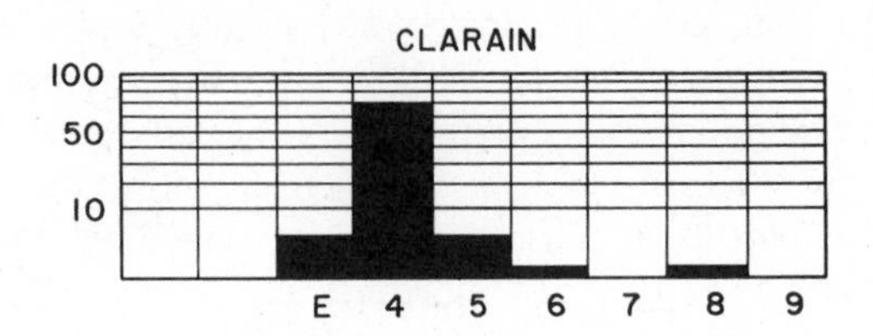

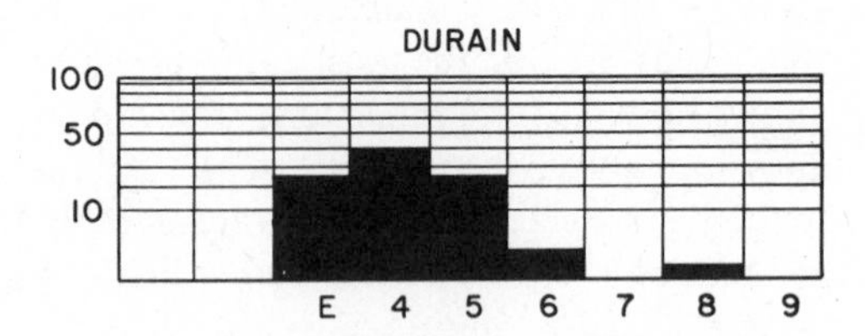

FIGURE 2.1 Variation of reflectance for selected maceral groups according to the method of C. A. Seyler. (*Source:* Reference 122; reprinted with permission from W. Francis, *Coal: Its Formation and Composition,* Edward Arnold, London, 1961, p. 342.)

clarain consists primarily of vitrinite ($N_r = 4.5$) and about 5% exinite (E). The durain has much less vitrinite and more exinite. In these two samples, fusain has an N_r value of 6 and 8, respectively (122).

The identification of the relative amount of the three maceral groups may provide all the information needed for many applications. Besides its utility in tracing and correlating seams and in classifying coals by rank and type, the primary use of petrographic data is process control (123). A coal that is rich in liptinite is preferred for hydrogenation because it is rich in hydrogen and aliphatic components (see Table 2.6). Microscopic determination of relative amounts of this maceral group in various coal seams aids the acquisition of the most desirable raw material for a hydrogenation process. Likewise, this high hydrogen content gives liptinite a higher heat content than the hydrogen-deficient vitrinites and inertinites. Bituminous coal rich in liptinite and vitrinite appears to be the most desirable for liquefaction. Liptinite and lesser amounts of vitrinite dissolve in the solvent oils, while most inertinites remain insoluble. Obviously, petrographic analysis is useful for determining the maceral ratios for coal destined for liquefaction. Liptinite is the least dense maceral group and has the lowest R_0 value. It appears dark gray. Inertinite is the densest maceral group and has the highest reflectance value. It appears white to light gray. Recall that inertinite has the lowest hydrogen content, lowest aliphatic content, and lowest volatile matter content. Vitrinite, on the other hand, appears intermediate gray. This coloring follows from the fact that its density and its reflectance are also intermediate between those of liptinite and those of inertinite. It also contains intermediate amounts of hydrogen, volatile matter, and aliphatic material.

For the preparation of coke, not only rank but also the plastic properties and swelling properties of coal are extremely important. These properties vary with the maceral composition of coal (124). Medium-volatile bituminous coal is generally used for coke. It is particularly rich in vitrinite, which swells and becomes plastic during the coking process. Rather than using coal of just one rank, most processes now rely on blends of coal to achieve a mixture of desirable qualities. It has been found that although inertinite alone is not suitable for preparing coke, it is a necessary component of a good coke. Apparently, it holds the coke mass together and gives it strength. Figure 2.2 shows how reflectance data can be used to analyze coal blends for their coking ability. The average percent volatile matter in each of the five blends is approximately the same, but the swelling index, which indicates the relative quality of the coke, decreases from 9 to 1. The reflectograms of the vitrinite content in each

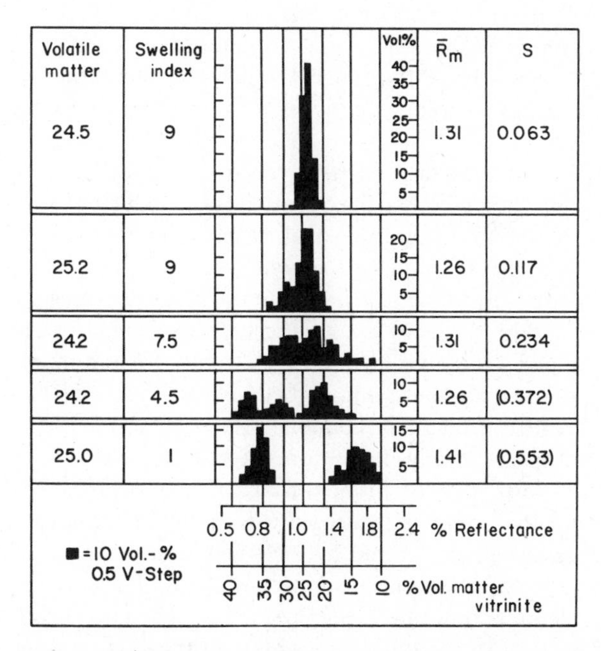

FIGURE 2.2 Application of reflectance data to determine coking character of various blends of coal. (*Source:* Reference 125; reprinted with permission from E. Stach et al., *Stach's Textbook of Coal Petrology,* Gebruder Bornfraeger, Stuttgart, 1982, p. 329.)

blend are shown in the center of the figure. It is obvious from these data that the best coking blends contain vitrinite with a volatile matter content between 20% and 30%, while the poorer-quality blends contain vitrinite for which the volatile content is either greater than 30% or less than 20%. By themselves, these coals would all yield a very poor coke, if any. Petrographic analyses are also valuable in the formulation of a "synthetic" coke called formed coke. Formed coke is a carbonized briquette made by using finely crushed coal mixed with noncoal binders. The advantages of formed coke is that it can be prepared from low-rank coal that would be unsuitable for normal coke production. Petrography is used to analyze various briquette mixtures and various preparation conditions in order to optimize the production of the highest-quality material. Petrographic analyses are also used to predict the yield of products, coke, tar, gas, and light oil that may be formed during carbonization reactions (126).

In addition to the use of microscopic analyses to monitor the final blends of coal going into a process, such analyses can also be used to adjust the preliminary mechanical process of sizing and screening coal before it enters the blend. It has been shown for Illinois Number 6 coal that, although the average vitrain content is 20%, the vitrain content rises as the coal is crushed and is about 79% when the coal is pulverized (127). It is apparent that vitrain becomes segregated as the screening process separates smaller and smaller sizes, and becomes most concentrated in the fines. Sizing, crushing, and screening may also affect the distribution of minerals in a particular fraction. Microscopic examination is used to determine this type of mechanical segregation as well. Similarly, microscopic determination of minerals as they occur in seams is used to construct "horizon markers" for seam correlation (128).

Paleobotanical studies of coal also rely on microscopic analyses. Some investigators claim that coal has been treated too much from a botanical point of view, but the extensive literature dealing with the botanical significance of macerals is very valuable to coal geologists and coal chemists alike. We have referred repeatedly to the plant residues in coal: spore exines, algae, cuticles, fungi, and bark. The examination of these constituents, especially of spore exines, from a paleobotanical perspective is used for taxonomic and stratigraphic studies aimed at identifying and classifying the forms of these entities. The specific identification of nonbanded coal—of whether it is boghead or cannel coal—is based strictly on the presence of algae and spore exine content. The characterization of unique coal, such as Moscow paper coal, is based largely on the botanical examination of cuticle remains of leaves that have survived essentially unaltered by their geologic history (129).

The applications of petrographical methods should also be mentioned with regard to petroleum and natural gas exploration. The vitrinite content of coal is a direct indication of the degree to which all the surrounding sedimentary rock has been altered by subsidence, tectonic movement, and geothermal gradients. Much work in this field has been conducted by M. and R. Teichmüller

(130). The investigation of oil shale is also being pursued on the basis of petrographic analysis. Jacobs and Alpern are two pioneers in this area of research (130). Likewise, soil science is beginning to make extensive use of in situ microscopic analysis instead of using time-consuming extraction processes to isolate the organic constituents (130). These few examples leave no doubt that microscopic studies of the entire coalification process will be increasingly used in all areas of coal handling, analysis, and technical processing, and that these same methods and techniques will also be increasingly used in fields of study related indirectly to coal.

2.11 INORGANIC CONSTITUENTS OF COAL

The mineral content of coal is important because it affects the performance, and therefore the value, of coal in several ways. One consideration is the relationship of mineral matter to the overall generation of heat energy produced when coal is burned. Although it is the combustion of organic hydrocarbons in coal that releases heat, the minerals present consume heat during the combustion process. Thus, the gross calorific value of a coal is affected adversely by the presence of mineral matter. Figure 2.3 illustrates how the heat content varies with the ash yield as well as with the volatile matter content. It may be critical to a coal producer to reduce the mineral content of a coal to provide a product with a higher market value than the coal normally produced by a given mine. In addition, minerals cause solid buildup and corrosion on boiler surfaces, due principally to alkali metals and alkali chlorides from clay minerals. These elements react with sulfides to form a variety of sulfate complexes. Bituminous coal is usually low in these species, but their presence in lower-rank coals may require costly pretreatment (132).

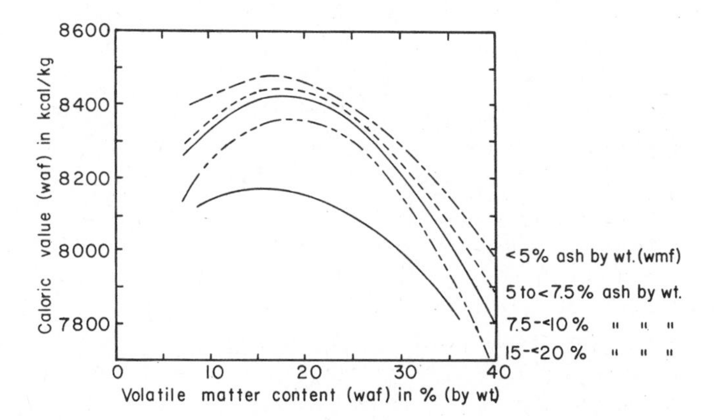

FIGURE 2.3 Relationship between calorific value and volatile matter content in coal of varying mineral matter content. (*Source:* Reference 131; reprinted with permission from E. Stach et al., *Stach's Textbook of Coal Petrology,* Gebruder Bornfraeger, Stuttgart, 1982, p. 467).

Spontaneous combustion is a perennial danger in mines and in stationary or in-transit piles of coal. The nature of these reactions is not completely understood, but it is generally agreed that the danger increases as weathering or air oxidation of finely dispersed pyrite produces sulfuric acid. When a dry coal is wetted, the probability of ignition is increased tenfold in the presence of this iron compound. Coal with a high vitrinite content is particularly susceptible to the occurrence of this reaction (133). The presence of pyrite is also of critical concern in coal processing. High-sulfur coals are undesirable for coke production because the sulfur contaminates the coke. Also, the conversion of pyritic sulfur to SO_2, released into the atmosphere principally by power plants that consume large amounts of coal, presents a very serious pollution problem. Another aspect of pollution related to mineral matter occurs in the mines. Finely dispersed quartz, oxides of silica, constitute "mine dust" and cause silicosis. Finally, just the buildup of waste ash from any coking or other carbonization process poses a problem of waste disposal. Much research is underway to find uses for the ash, such as combining it with tars to form a road pavement. It was shown earlier in Section 2.3 how a variety of metal cations and acid anions can become incorporated in coal during biogenesis of peat. Minerals are found in all types of peat. Because lowmoor peat bogs have more contact with groundwater, they tend to contain a more varied array of mineral matter (134).

Inorganic components in coal are classified as primary vegetable ash or as secondary minerals. Primary vegetable ash minerals are those inherent in plant matter that formed coal. They are usually dispersed intimately and generally throughout the coal but are not very abundant. Secondary minerals are of two types: those that are extraneous to the plant matter but were deposited by wind and water during diagenesis and may be chemically bound to organic structures, and those that accumulated during the second stage of coalification primarily by the percolation of water through the seams. The first type is called syngenetic mineral matter, and the second type is called epigenetic mineral matter (135). Like inherent minerals, syngenetic minerals are usually fine grained and widely distributed and can be used as horizon markers for seam identification. The concentration of extraneous minerals almost always exceeds that of the inherent inorganic content of coal and is of the most interest in coal technology. Epigenetic minerals tend to be neither finely dispersed nor intimately ingrown with the coal. They occur randomly in a seam and can generally be separated from the mined coal by physical methods such as crushing and sieving, or by a washing technique called float-sink separation, which makes use of the difference in density between minerals and coal when the solids are dispersed in a suitable medium. Magnetic methods and electrostatic separation techniques are also used to clean coal. Table 2.9 gives a survey of the minerals found in coal and a description of their origin.

Although these distinctions are clear in principle, it is not always possible to clearly assign a mineral component to one class or another. Colloids, for

TABLE 2.9 Summary of the Minerals in Coal

	First Stage of Coalification: Syngenetic Formation, Synsedimentary, Early Diagenetic (intimately intergrown)		Second Stage of Coalification: Epigenetic Formation	
Mineral Group	Transported by Water or Wind	Newly Formed	Deposited in Fissures, Cleats, and Cavities (coarsely intergrown)	Transformation of Syngenetic Minerals (intimately intergrown)
Clay minerals	Kaolinite, illite, sericite, clay minerals with mixed-layer structure, montmorillonite, tonstein			Illite, chlorite
Carbonates		Siderite-ankerite concretions, dolomite, calcite, ankerite, siderite, calcite	Ankerite, calcite, dolomite, ankerite in fusite	
Sulphides		Pyrite concretions, melnikovite-pyrite, coarse pyrite (marcasite), concretions of FeS_2—$CuFeS_2$—ZnS	Pyrite, marcasite, zinc sulphide (sphalerite), lead sulphide (galena), copper sulphide (chalcopyrite), pyrite in fusite	Pyrite from the transformation of syngenetic concretions of $FeCO_3$
Oxides		Hematite	Geothite, lepidocrocite ("needle iron ore")	
Quartz	Quartz grains	Chalcedony and quartz from the weathering of feldspar and mica	Quartz	
Phosphates	Apatite	Phosphorite, apatite		
Heavy minerals and accessory minerals	Zircon, rutile, tourmaline, orthoclase, biotite		Chlorides, sulphates, and nitrates	

Source: Reference 136; E. Stach et al., *Stach's Textbook of Coal Petrology,* Gebruder Bornfraeger, Stuttgart, 1982, p. 165.

instance, which may contain inorganics, are not truly chemical combinations of elements. Also, some material that accumulated by percolation may be as finely grained and as generally dispersed as syngenetic material. Finally, there is no clear separation between elements that can definitely be associated with plant metabolism and those that definitely are never found in plants (137).

Microscopic analysis of the inorganic constituents of coal can be performed with either thin sections using Thiessen's grinding technique or with polished sections. Stopes and Wheeler were among the first to investigate the inorganic content of lithotypes and to show that the ash content increased in the following order: vitrain, clarain, durain, and fusain. It seems clear that the ash in vitrain is mainly from plants, while that in durain is from clay washed into accumulating peat deposits. The composition of inorganics in clarain is intermediate between these two. The nature of the inorganic matter in fusain is variable, but fusain always contains the highest quantity (138).

In addition to microscopic analysis, it is also possible to analyze the inorganic components of coal by spectrometric methods. For the analysis of inorganic components, atomic absorption spectroscopy has proved immensely valuable. Nearly every metal found in coal can be determined by this method. Spectrophotometric determination of sulfur has many advantages over the older "wet" technique. Other methods, such as x-ray diffraction and x-ray fluorescence, scanning electron microscopy, and infrared and emission spectroscopy, are all described in Chapter 5.

Clay minerals are by far the most abundant inorganic constituents of coal and may constitute as much as 80% of the total mineral content. All coal and microlithotypes can contain minerals, and three classes of clay content are identified by microscopic analysis: less than 20% clay by volume is considered contaminated, 20–60% clay by volume is termed carbargilite, and more than 60% clay by volume is called argillaceous shale (139). Clay minerals are important in mining and in processing. All clay minerals swell when contacted with water, and the particle aggregates tend to disintegrate as swelling occurs. In mines, this combination of wetting and disintegrating causes floors to buckle and roofs to collapse. In processing, the effect is to produce quantities of slime, which makes dewatering difficult. A high clay content also increases the ash fusion temperature of the coal. This is an important factor for boiler fuels (see Chapter 5 for a discussion of ash fusion). Clay, of course, is also a refractory material, and if it occurs in large layers, its recovery is economically practical.

Carbonates are the second most frequent mineral group to occur in coal. The presence of dolomitic limestone is evidence of marine intrusion during diagenesis. Dolomite [$CaMg(CO_3)_2$] is the principal inorganic component of coal balls, aggregates of completely preserved plant remains. *Carbankerite* is the term applied to microlithotypes containing 20–60% by volume of carbonates.

Although, compared to clay minerals and carbonates, most sulfide material constitutes only a small percentage of the total mineral content of coal (Table

2.9), we have referred to sulfur often as being a mineral that evokes considerable interest for coal utilization. Coal from the Interior Province of the United States contains a relatively high sulfur content. The sulfides rarely account for even 5% of the mineral content of coal, but most of the adverse impact on coal handling and use described above is associated with pyrite. Microbial decomposition of organic sulfur-containing compounds and the formation of pyrite and marcasite were described earlier (section 2.3). The reduction of sulfides to elemental sulfur occurs infrequently. Oxidation of sulfur results in the formation of sulfur dioxide, which in an aqueous environment can form water-soluble sulfates (140). This is the process that produces acid mine drainage and, at least in part, acid rain.

Oxides in the form of quartz were also mentioned earlier in this section. No other oxide or hydroxide is of special interest with regard to coal. Iron ore may occur in relatively large layers near coal seams that have unexpectedly high rank. In Germany particularly, these layers frequently form outcrops and have given rise to the iron and steel industries in that country (141).

Phosphates are of academic value in that they originate in protein albumin. As mineral matter, phosphates are especially undesirable in coke for electrometallurgical applications. Heavy minerals in coal have not been thoroughly investigated. It is known that some, such as zircon, are frequently associated with radioactive elements. Salts are always undesirable because of their extreme corrosiveness. Salt minerals may be intimately ingrown in coal but occur rarely in most American coal (118). In some German and English coal fields, the salt minerals are so abundant that coal is termed "salt coal" (141).

Trace elements in coal occur in about the same concentration as in the earth's crust: about 0.1% or less. Systematic studies of trace elements is an area of current research. Many elements in this category—uranium, germanium, arsenic, boron, and beryllium—may occur in concentrations substantially greater than their average concentration in the earth's crust. Some others—barium, bismuth, cobalt, copper, gallium, lanthanum, lead, lithium, mercury, molybdenum, nickel, scandium, selenium, silver, strontium, tin, vanadium, yttrium, zinc, and zirconium—may also occur in coal in levels higher than those generally found. All these elements have become increasingly important as possible resources and as potential pollutants.

Some discussions of the composition or analysis of coal use the terms *ash* and *inorganic matter* interchangeably. This is not, strictly speaking, correct. Ash obtained by the combustion of coal differs in amount and nature from the inorganic components of the coal. For classification purposes the actual inorganic constitution should be known. To use ash content as an index of inorganics is inaccurate because chemical changes in inorganic components may be severe during combustion. These changes include loss of water (in clays), loss of carbon dioxide (in carbonates), oxidation (in sulfides), and decomposition or volatilization (in halides). If the coal cannot be freed of most inorganics before analysis, factors must be applied to the calculation of the percent ash in order to correct for these differences (142). These corrections are addressed in Chapter 5.

SUMMARY

This chapter has shown that coal was formed by a number of complex processes that occurred simultaneously and in overlapping sequences.

The combination of geological, chemical, and physical circumstances that coincided during distinct periods of time give evidence of being sufficiently unusual that they may never occur again. Although natural and simulated environments are used as models, our ability to reconstruct past events accurately is still quite limited.

Coal petrologers and coal petrographers use both simple and very sophisticated techniques to determine the botanical composition and the physical character of coal. The classifications of coal that result from these studies are very important for the most judicious use of our coal resources as well as for our basic knowledge of coal.

The organic chemistry of coal is the topic of Chapter 3, and the impact of petrographic analyses on coal utilization is presented in Chapter 4.

REFERENCES

1. W.T. Thom, Jr., *Petroleum and Coal: The Keys to the Future,* Princeton University Press, Princeton, NJ, 1929, pp. 42–44.
2. L.R.M. Cocks, ed., *The Evolving Earth,* British Museum and University of Cambridge Press, 1981, London pp. 145–146.
3. E.S. Moore, *Coal: Its Properties, Analysis, Classification, Geology, Extraction, Uses and Distribution,* 2d ed., Wiley, New York, 1940, p. 350.
4. F.J. Rich, *Modern Wetlands and Their Potential as Coal Forming Environments* [abstr.]: Bulletin—Corpus Christi Geological Society, March 1983, p. 4.
5. Thom, p. 16.
6. D.N. Cargo and B.F. Mallory, *Man and His Geologic Environment,* Addison-Wesley, Reading, MA, 1974, p. 217.
7. Moore, pp. 146–147.
8. Moore, pp. 158–166.
9. V. Bouška, *Geochemistry of Coal,* Elsevier, Amsterdam, 1981, p. 38.
10. J. Taylor and R. Smith, "Power in the Peat Lands," *New Scientist, 88,* 644–646 (1980).
11. W.L. Helton, private communication 1982.
12. Cargo and Mallory, p. 7.
13. K. Lindberg and B. Provorse, *Coal: A Contemporary Energy Story,* R. Conte, ed., Scribe and *Coal Age,* New York, 1977, p. 14.
14. W. Francis, *Coal: Its Formation and Composition,* Edward Arnold, London, 1961, pp. 188–189.
15. Bouška, pp. 16–20.

16. O. Stutzer, *Geology of Coal,* A.C. Noé, trans., University of Chicago Press, Chicago, 1940, p. 90.
17. Bouška, pp. 73–74.
18. D. Casagrande and K. Siefert, "Origins of Sulfur in Coal: Importance of the Ester Sulfate Content of Peat," *Science, 195,* 675–676 (1977), and reference therein.
19. Francis, pp. 6–9; Moore, p. 138.
20. Moore, pp. 136–138.
21. T.A. Hendricks, "The Origin of Coal," in H.H. Lowry, ed., *Chemistry of Coal Utilization,* vol. 1, Wiley, New York, 1945, pp. 2–8.
22. Francis, pp. 4–29.
23. Moore, pp. 142–143.
24. Francis, pp. 35–36.
25. Moore, p. 151.
26. Stutzer, pp. 88–89.
27. Stutzer, p. 90.
28. E. Stach, G.H. Taylor, M.-Th. Mackowsky, D. Chandra, M. Teichmüller, and R. Teichmüller, *Stach's Textbook of Coal Petrology,* D.G. Murchison, G.H. Taylor, and F. Zierke, trans., 3d ed., Gebruder Bornfraeger, Stuttgart, 1982, pp. 27–28.
29. Francis, pp. 223–224, 629.
30. Stach et al., pp. 31–33.
31. Francis, pp. 155–157, 208.
32. D.W. Van Krevelen, *Coal,* Elsevier, Amsterdam, 1961, p. 440.
33. Bouška, pp. 81–83.
34. I.A. Breger, "Geochemistry of Coal," *Economic Geology, 53,* 823–826 (1958), Stutzer, pp. 99–103; Hendricks, pp. 11–12.
35. D. Muchison and T.S. Westoll, eds., *Coal and Coal-Bearing Strata,* American Elsevier, New York, 1968, p. 199.
36. Moore, p. 159.
37. Stach et al., pp. 17–19.
38. R.A. Schmidt, *Coal in America: An Encyclopedia of Reserves, Production, and Use, Coal Week* and McGraw-Hill, New York, 1979, pp. 14–15.
39. Francis, pp. 216–220, 567.
40. See especially the more recent texts by Stach et al. and Bouska.
41. Hendricks, p. 9.
42. Bouška, p. 97; Stach et al., p. 33.
43. Stach et al., pp. 42–43.
44. Stach et al., p. 55.
45. Thom, p. 29.
46. Bouška, p. 126.
47. Francis, p. 219.
48. Breger, pp. 826–827.
49. Stach et al., pp. 34, 38–39, 60.
50. Stach et al., p. 38.
51. Moore, pp. 94–95.
52. Bouška, p. 103.
53. Schmidt, p. 20.

54. Breger, pp. 827–828.
55. Francis, pp. 568–569.
56. Hendricks, p. 14.
57. Francis, pp. 572–575.
58. Bouška, p. 108.
59. Francis, p. 610.
60. Francis, pp. 608–611; M. Teichmüller, *Coll. Int. Pet., Rev. de l'Ind. Min.*, Num. Spec., 99 (1958).
61. Stutzer, p. 292.
62. Stach et al., pp. 55, 59, 63, 64.
63. Bouška, p. 105.
64. Stutzer, pp. 293–299.
65. Stutzer, p. 294.
66. Francis, p. 575.
67. Stutzer, p. 297.
68. *The Story of Anthracite,* Hudson Coal, New York, 1932, p. 5.
69. Stach et al., p. 58.
70. Bouška, p. 100.
71. M. Teichmüller and R. Teichmüller, *Z. Neutsch. Geol. Ges., 117,* 243–279 (1966), cited in Stach et al., p. 58.
72. Moore, p. 181.
73. A. Raistrick and C.E. Marshall, *The Nature and Origin of Coal and Coal Seams,* University Press, London, 1952, pp. 251, 255–256.
74. Francis, p. 585.
75. Raistrick and Marshall, pp. 253–254.
76. Stutzer, p. 299.
77. Moore, p. 179.
78. R.H. Severson, "Depositional Environments, Facies Relationships and Coal Occurrence in Carboniferous Sediments of the Narragansett Basin," M.S. thesis, University of Rhode Island, Kingston, 1981, pp. 1, 9–14.
79. R.A. Mott, *Fuel, 21,* 129 (1942); *22,* 20 (1943).
80. J.P. Schümmacher, F.J. Huntjens, and D.W. Van Krevelen, *Fuel, 14,* 223 (1960).
81. Francis, pp. 438–449.
82. D.J. Cuff and W.J. Young, *The United States Energy Atlas,* Free Press, New York, 1980, pp. 70–71.
83. Francis, pp. 602–604.
84. H.J. Rose, "Classification of Coal," in H.H. Lowry, ed., *Chemistry of Coal Utilization,* vol. 1, Wiley, New York, 1945, pp. 25–85.
85. Stach et al., p. 1.
86. F. Link, *Abhandl. k. Preuss Akad. Wiss. Berlin, 34,* 33–34 (1838); F. Schulze, *Ber. k. Akad. Wiss. Berlin,* 676–678 (1855); cited in Moore, p. 11.
87. Van Krevelen, pp. 483–494.
88. Stach et al., pp. 1–4.
89. R. Thiessen, *Coal Age, 18,* 1183 (1920); *J. Geol., 28,* 185 (1920).
90. Francis, pp. 330–331.
91. R. Thiessen, *Methods and Apparatus Used in Determining the Gas-, Coke-, and by-product-Making Properties of American Coals, with Results on a Taggart Bed Coal from Roda, Wise*

County, Va, U.S. Bureau of Mines Bulletin no. B 344, Pittsburgh, 1931, p. 64; D. White, *The Effect of Oxygen in Coal,* U.S. Bureau of Mines Bulletin no. B 29, Pittsburgh, 1911.

92. Stach et al., p. 173.
93. W. Spackman, A. Davis, and G.D. Mitchell, "The Fluorescence of Liptinite Macerals," *Brigham Young Univ. Geol. Studies, 22*(part 3), 60–61 (1976).
94. M.C. Stopes, "On the Four Visible Ingredients in Banded Bituminous Coal," *Proc. Roy. Soc., 90B,* 470 (1919); *Fuel, 14,* 6 (1935).
95. Moore, p. 23.
96. Francis, p. 334.
97. D. White and R. Thiessen, *The Origin of Coal,* U.S. Bureau Mines Bulletin no. 38, Pittsburgh, 1913, cited in Hendricks, p. 1.
98. L. Petrakis and D.W. Grandy, "Coal Analysis, Characterization and Petrography," *J. Chem. Ed., 57,* 689–694 (1980).
99. G.H. Cady, "Coal Petrography," in H.H. Lowry, ed., *Chemistry of Coal Utilization,* vol. 1, Wiley, New York, pp. 86–131.
100. W. Spackman, "The Maceral Concept and the Study of Modern Environments as a Means of Understanding the Nature of Coal," *Transactions of the New York Academy of Science,* series 2, *20*(5), 411–423 (1958).
101. M.R. Campbell, *Geology of the Big Stone Gap Coal Field of Virginia and Kentucky,* United States Geological Survey Bulletin no. 111-B, 1893.
102. "Megascopic Description of Coal and Coal Beds and Microscopical Description and Analysis of Coal," in *Annual Book of ASTM Standards,* vol. 05.05, section D2796, American Society for Testing and Materials, Philadelphia, 1984.
103. Moore, pp. 11–16; Cady, p. 105.
104. E.C. Jeffrey, "The Nature of Some Supposed Algal Coals," *Proc. Amer. Acad. Arts and Sci., 46,* 273–390 (1910); E.C. Jeffrey, "Methods of Studying Coal," *Conspectus, 6*(3), 1916; cited in Moore, p. 13.
105. Stach et al., p. 299.
106. D. White and R. Thiessen, *The Origin of Coal,* U.S. Bureau of Mines Bulletin no. 38, Pittsburgh, 1913, p. 207, cited in Moore, p. 14.
107. White and Thiessen, pp. 211–216.
108. Stach et al., pp. 296–299.
109. Moore, p. 18.
110. Van Krevelen, p. 103; Moore, p. 18.
111. "Preparing Coal Samples for Microscopical Analysis by Reflected Light," in *Annual Book of ASTM Standards,* vol. 05.05, section D2797, American Society for Testing and Materials Philadelphia, 1984.
112. Stach et al., p. 2.
113. Cady, pp. 103, 113.
114. Stach et al., pp. 306–311.
115. G.J. Jansen, "The Petrography of Western Coals," *Proceedings of the Second Symposium on the Geology of Rocky Mountain Coal,* H.E. Hodgson, ed., Colorado Geological Survey, Denver, 1978, p. 182.
116. Stach et al., p. 55.
117. Petrakis and Grandy, pp. 691–692.
118. Stach et al., p. 4.
119. Petrakis and Grandy, p. 693; Stach et al., p. 142.
120. H. Tschamler and E. deRuiter, *Coal Science,* R.F. Gould, ed., American Chemical Society, Washington, 1966, p. 333.

121. A. Davis, *The Measurement of Reflectance of Coal Macerals: Its Automation and Significance,* Department of Energy Report no. DOE-FE-2030-TR10, Washington, June 1978, p. 68, *Energy Abstr.* no. 35591, 1979.
122. Francis, p. 342.
123. Stach et al., pp. 413–423; Francis, pp. 437–438.
124. Stach et al., pp. 424–434.
125. Stach et al., p. 329.
126. Francis, p. 306; Cady, pp. 127–130; Stach et al., pp. 423–424, 455–466.
127. Cady, p. 131.
128. Petrakis and Grandy, p. 693; Stach et al., pp. 375–376.
129. Stach et al., pp. 115, 252.
130. Stach et al., pp. 373, 374, 399–412.
131. Stach et al., p. 467.
132. Stach et al., pp. 468–474.
133. Stach et al., pp. 474–476.
134. Bouška, p. 92.
135. Petrakis, p. 690; Stach et al., p. 193; Francis, pp. 630, 639.
136. Stach et al., p. 165.
137. Francis, p. 636.
138. Francis, p. 646.
139. Stach et al., pp. 158–159.
140. Bouška, p. 95.
141. Stach et al., p. 399.
142. Francis, p. 659.

3
THE ORGANIC STRUCTURE OF COAL

One of the most challenging problems facing the organic chemist today is the determination of the chemical structure of coal. The complexity and heterogeneity of coal make it impossible to determine the structure of a "coal molecule." It is possible, however, to determine the structural types that are most frequently encountered, some of the changes in structure that accompany changes in coal rank, and many of the basic chemical reactions that occur during combustion and the conversion of coal to liquid and gaseous fuels. In so doing, we can learn much more about the biochemical and metamorphic changes that occur during the coalification process. This knowledge should ultimately allow us to achieve a more efficient utilization of coal as an energy source and of conversion to clean liquid and gaseous fuels.

Coal research was actually first developed by geologists. This work followed rather closely the demand for coal and the technical development of coal usage by industry. This applied research was followed by what is often called the classical period of coal research. The bulk of this work was done between the turn of the century and the Second World War. This research dealt largely with the pyrolysis, solvent extraction, oxidation, and reduction of coal.

One of the major problems encountered in the determination of the chemical structure of coal is the heterogeneity of the samples. The chemical composition of coal not only varies from seam to seam, but rather large variations can also occur within a single seam. Indeed, there is also variation among the different macerals within a single lump of coal. Another problem is the size and complexity of the molecules involved. The methods employed are quite different, but in many respects the problems associated with the determination of the structure of coal closely parallel those encountered in the determination of protein structure.

In this chapter we shall first study the classical reactions of coal in which the coal matrix is broken down into smaller molecules. This information, coupled with information derived from instrumental studies on whole coal, will be used to develop some ideas about the nature of organic structures and bonding in coal. Once we have a simple understanding of the nature of the bonding and the organic structures present in coals of different rank, we shall use this information in Chapter 4 in an attempt to understand some of the basic processes occurring during the more common coal conversion techniques.

3.1 COAL PYROLYSIS

One of the very early chemical reactions of coal to be investigated was pyrolysis, or "carbonization." Pyrolysis is defined as the decomposition, or breakdown, of a substance by means of heat. This reaction has not only been used to study the structure of coal, but is also widely used in the manufacture of coke and of a variety of organic chemicals. While pyrolysis is admittedly a rather crude tool, the results obtained from this procedure can make significant contributions to the determination of the organic structure of coal. The similarity of the volatile products from low-temperature ($< 500°C$) pyrolysis to the original coal implies that the products are most likely derived from parts of the coal structure. Analysis of these products would, then, provide valuable information about the chemical structure of coal itself.

Mass spectrometric and other data indicate that the major components of the solid, liquid, and gaseous products of the low-temperature pyrolysis of coal have molecular weights below 400 amu. The complexity of the problem one faces in the determination of coal structure via pyrolysis can be appreciated on learning that well over 800 different compounds have been isolated and characterized as products from "coal tar." One must also realize that the majority of the compounds are obtained in yields of less than 1%.

The most widely held view of the nature of the mechanism of coal pyrolysis involves cleavage of one or more covalent bonds holding the organic fragment to the coal matrix. A covalent bond may be broken in two ways, yielding either free radicals or ions:

Homolytic cleavage

$$A{:}B \longrightarrow A^{\cdot} + B^{\cdot}$$

Heterolytic cleavage

$$A{:}B \longrightarrow A^{+} + B{:}^{-}$$

In the case of homolytic bond cleavage, each atom (fragment) retains one electron of the shared electron pair that originally made up the covalent bond. These atoms or molecular fragments with an unpaired electron are called free radicals. In the case of heterolytic bond cleavage, one atom (fragment) retains both electrons of the shared pair. This forms a cation and an anion. Most chemists agree that the bulk of coal pyrolysis reactions are initiated by the thermal cleavage of a few covalent chemical bonds within the coal matrix. This gives rise to free radicals, which can then further react in a number of different ways. The radical may abstract hydrogen from an internal or external source to yield a stable molecule. It may fragment into smaller radicals, which are in turn stabilized by reaction with hydrogen. The radicals may also disproportionate (one radical abstracting a hydrogen from another) to yield stable molecules. The smaller molecules formed would most likely be volatile and distill from the coal. Alternatively, the free radicals may couple (combine) to form char or coke. In addition, the hydroaromatic structures present in the coal matrix may undergo dehydrogenation to yield an extended aromatic system that remains in the char or coke. A highly simplified mechanism of some of the reactions that may occur in coal pyrolysis is shown in Figure 3.1.

The results of research by S. R. Gun and coworkers strongly supports the free radical view of the coal pyrolysis mechanism (1). Table 3.1 summarizes this work. The radical scavengers benzophenone and iodine are both effective in reduction of the yield of gaseous and benzene-soluble products. If species that are known to react rapidly with free radicals (radical scavengers) affect the rate of a reaction and overall product yield, one may conclude that free radicals are involved. The results in Table 3.1 also show the effect of added hydrogen on stabilization of the radicals before coupling to form char or coke. The addition of hydrogen (from an external source) results in an increase in the yield of volatile matter. Replacement of the inert nitrogen atmosphere by hydrogen increases the yield of volatile products from 16.3% of the dry, ash-free weight of the coal to 46.6% of its dry, ash-free weight. Using a combination of a hydrogen atmosphere for the pyrolysis and an ammonium

1. Thermolysis of bond within the coal matrix

$$R{:}R \xrightarrow{\text{heat}} R^{\cdot} + R^{\cdot}$$

2. Addition of hydrogen from internal or external source

$$R^{\cdot} + H^{\cdot} \longrightarrow R{:}H$$

3. Fragmentation

$$R^{\cdot} \longrightarrow R'{:}H + R''^{\cdot}$$

4. Disproportionation

$$2R^{\cdot} \longrightarrow R{:}H + R'{-}CH{=}CH_2$$

4. Coupling to form coke or char

$$R^{\cdot} + R^{\cdot} \longrightarrow R{:}R$$

FIGURE 3.1 Simplified coal pyrolysis mechanism.

TABLE 3.1 Pyrolytic Conversion of Coal to Gases and Benzene-Soluble Products

Gas Pressure	Catalyst (wt% coal)	Radical Scavenger (wt% coal)	Coal Conversion (wt% coal)
9.8 MPa Nitrogen	—	—	16.3
9.8 MPa Nitrogen	—	22% Benzophenone	5.7
9.8 MPa Nitrogen	—	10% Iodine	5.5
9.8 MPa Nitrogen	0.1	—	16.9
9.8 MPa Hydrogen	—	—	46.6
9.8 MPa Hydrogen	—	22% Benzophenone	21.6
9.8 MPa Hydrogen	0.1	—	75.3

Source: Reference 1; reprinted with permission from S.R. Gun et al., "A Mechanistic Study of Hydrogenation of Coal, Part 2," *Fuel, 58,* 176, 1979, Butterworth & Co., Publishers, Ltd.

Note: Coal used was 79.9% C, 5.9% H, 4.4% S, 1.3% N, 8.5% O, and 43.8 % volatile matter; 25 g of coal were heated to 400°C for 1 h. Catalyst was ammonium molybdate impregnated in the coal. MPa is pressure in Megapascals. A pressure of 1 standard atmosphere is approximately equal to 0.1 MPa.

molybdate catalyst increased the yield of volatile products further to 75.3%. As we shall see in Section 4.4, these are the approximate conditions for the hydroliquefaction of coal.

Inspection of Table 3.2 and studies of model compounds suggest that the C—O bond of benzylic ethers (compounds of the type Ar—CH_2—O—Ar, Ar—CH_2—O—CH_2—Ar, or Ar—CH_2—O—R, where Ar is an aromatic ring and R is an alkyl group) and the C—C bond of ethylene bridges connecting aryl groups (Ar—CH_2—CH_2—Ar) are very good candidates for low-temperature scission (2,3). Benzyl phenyl ether yields toluene (C_6H_5—CH_3) and phenol (C_6H_5—OH), while 1,2-diphenylethane yields only toluene on pyrolysis at 400° C in the presence of a hydrogen source:

$$C_6H_5\text{—}CH_2\text{—}O\text{—}C_6H_5 \xrightarrow{400^\circ C} C_6H_5\text{—}\dot{C}H_2 + C_6H_5\text{—}\dot{O}$$

$$C_6H_5\text{—}\dot{C}H_2 + C_6H_5\text{—}\dot{O} \xrightarrow{[H]} C_6H_5\text{—}CH_3 + C_6H_5\text{—}OH$$

$$C_6H_5\text{—}CH_2\text{—}CH_2\text{—}C_6H_5 \xrightarrow{400^\circ C} 2\ C_6H_5\text{—}\dot{C}H_2$$

$$2\ C_6H_5\text{—}\dot{C}H_2 \xrightarrow{[H]} 2\ C_6H_5\text{—}CH_3$$

TABLE 3.2 Some Representative Bond Dissociation Energies

Compound	Bond Dissociation Energy (kcal/mole)	Half-life ($t_{\frac{1}{2}}$) at 450° C[a,b]
$C_6H_5-C_6H_5$	103	301.5×10^6 yr
$RCH_2CH_2-CH_2CH_2R$	83	271.1 yr
$C_6H_5-CH_2C_6H_5$	81	67.4 yr
RCH_2-OCH_2R	80	33.6 yr
$C_6H_5CH_2-CH_3$	72	46.8 da
$C_6H_5CH_2-CH_2CH_2C_6H_5$	69	5.8 da
$C_6H_5CH_2-OCH_3$	66	17.2 h
$C_6H_5CH_2-CH_2C_6H_5$	57	2.0 min
$C_6H_5CH_2-OCH_2C_6H_5$	56	58.9 s
$C_6H_5CH_2-OC_6H_5$	51	1.8 s
$C_6H_5CH_2-SCH_3$	51	1.8 s
$CH_2=CHCH_2-CH_2CH=CH_2$	38	2.1×10^{-4} s

Note: C_6H_5 is the phenyl group; R is an alkyl group.
[a] $k = 10^{15} e^{-BDE/RT}$, where BDE is the bond dissociation energy, R the gas constant, and T the absolute temperature.
[b] $t_{\frac{1}{2}} = 0.693/k$.

The covalent bonds in the benzylic ethers and similar compounds are relatively easily broken because free radicals are being formed that are "resonance stabilized." Resonance is said to exist when more than one correct structural formula can be drawn for a molecule, ion, or radical. The multiple structures illustrate the fact that the odd electron is not isolated on a single atom in the radical but is "delocalized" over the entire structure. Molecules, ions, and radicals that contain delocalized electrons (not isolated on a single atom) are much more stable than they would otherwise be.

Benzylic radical

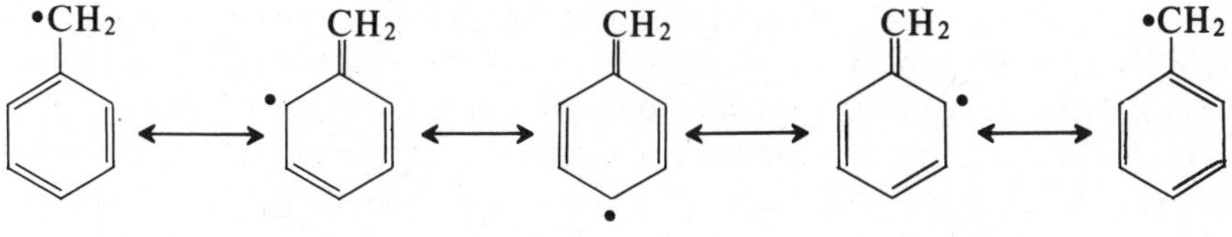

•CH2 ⟷ CH2 ⟷ CH2 ⟷ CH2 ⟷ •CH2

Phenoxy radical

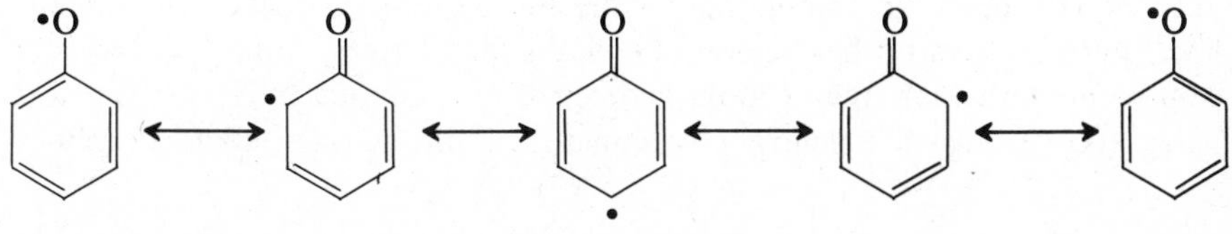

•O ⟷ O ⟷ O ⟷ O ⟷ •O

Substituents on the aromatic ring may have a profound influence on the thermal stability of a compound. Diphenylmethane (Ar—CH_2—Ar), for example, is rather stable at 450°C, with a half-life of approximately 67 years (Table 3.2). D. F. McMillen and coworkers have found that a hydroxyl group (—OH) attached to the aromatic ring renders the compound quite reactive at 400° (4):

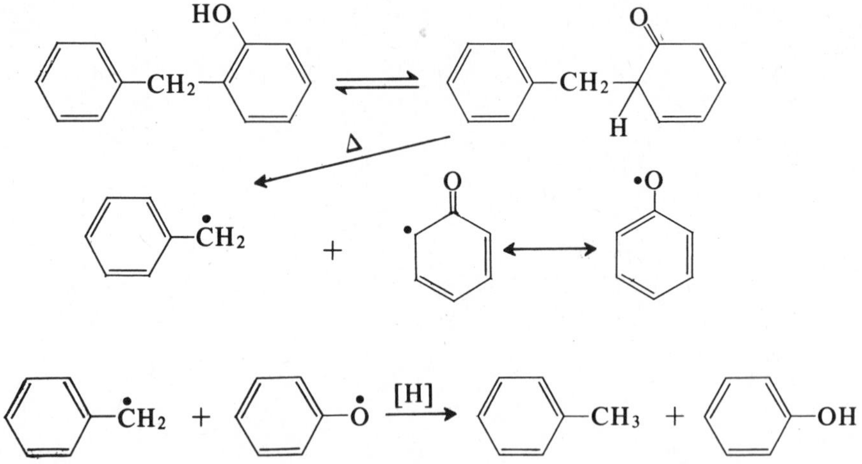

HO CH2 ⇌ O CH2 H

Δ

•CH2 + O • ⟷ •O

•CH2 + •O [H]→ CH3 + OH

One must consider the fact that coal is not just a mixture of hydrocarbons but contains a variety of organic functional groups such as hydroxyl. These groups will affect the thermal behavior of the coal.

The actual nature of the products obtained in a coal pyrolysis experiment depends on such factors as the rank of the coal being pyrolyzed, the pyrolysis temperature, and the contact time of the volatile matter with the char or coke and heat. Even different macerals from coal of the same rank show distinctly different behavior when subjected to pyrolytic decomposition. The volatile products from exinite, for example, are characterized by lower percentages of phenolic materials and higher yields of neutral compounds than are normally obtained from the corresponding vitrinite. In addition, gas chromatography and infrared analysis of the pyrolysis products show clearly that exinite contains a higher percentage of alkanes than does the corresponding vitrinite. The alkanes from exinite are predominantly linear, consisting of straight-chain hydrocarbons rather than branched alkanes. The predominant material from vitrinite is aromatic in nature.

The low-temperature pyrolysis product (coal tar) can be divided into three broad classes on the basis of acid-base behavior. The acidic component contains mostly phenolic compounds, while the basic component contains anilines, pyridines, and other heterocyclic amine derivatives. Amines are organic compounds in which one or more hydrogens of ammonia (NH_3) are replaced by an alkyl group. The neutral component contains alkanes, alkenes, cycloalkanes, hydroaromatics, aromatics, and various neutral ether and heterocyclic derivatives. Aromatic compounds are those similar to benzene and are unusually stable due to electron delocalization. Hydroaromatic compounds contain the same basic carbon skeleton as aromatic compounds but have too many hydrogen atoms for aromaticity. Hydroaromatics tend to lose hydrogen readily to form an aromatic ring. Heterocyclic compounds are those compounds that contain a ring made up of more than one kind of atom.

The product distribution among the coal tar fractions is highly dependent on the nature of the coal being pyrolyzed. By analysis of the product distribution among the various types of compounds, one can gather valuable information about the kinds of organic material present in coals of various ranks. The pyrolysis of lignite, for example, yields a much higher proportion of alkanes, alkenes, and cycloalkanes than does a similar pyrolysis reaction of a higher-rank coal. This would almost certainly indicate that the lignite contains more nonaromatic material as structural components than does the higher-rank coal. It should also be noted that the aromatic component of lignite also contains less highly condensed material (typically one or two rings) than does the aromatic material from higher-rank coal. Studies with advanced instrumental techniques, such as ^{13}C nuclear magnetic resonance (NMR) spectroscopy indicate that coal tar from high-rank coals consists almost exclusively of condensed aromatic ring systems. When a number of aromatic rings are fused (share common carbon atoms), the resulting compounds are called polynuclear aromatic compounds. The aliphatic (nonaromatic) contributions to the coal tars from high-rank coals are largely due to methyl groups on aromatic rings and to hydroaromatic groups, as opposed to the linear alkanes, which are present in lignitic coal tar. This would certainly indicate that the coal matrix for the high-rank coal is characterized by more highly condensed aromatic ring systems with the majority of the aliphatic material present as substituents on the rings. Lignite, on the other hand, appears to have a much less highly condensed matrix with considerably larger amounts of aliphatic material also present.

Studies of coal pyrolysis at temperatures below 400°C have also yielded information about the overall structure of coal. It is generally conceded that the energy of activation for the processes occurring at the lower temperatures (150–350°C) is too small (5–22 kcal/mole) for breaking normal covalent bonds. It is more likely that the organic material evolved in this temperature range is due to the vaporization of organic substances trapped within the capillary system of the coal particle and perhaps also to the liberation of materials bound to the coal matrix by weaker interactions (e.g., hydrogen bonding). If this view is correct, the rate of evolution of the volatile products would depend on their size and shape, and on the pore size and structure of the coal being pyrolyzed. There is evidence that, at least to some extent, tar production is indeed controlled by diffusion through the pores in the coal particle (5,6). It has been observed that pretreatment of certain coals with solvents that cause swelling and enlargement of the pores also results in higher tar yields on pyrolysis (7). Vacuum pyrolysis studies (pyrolysis of coal under reduced pressure) and studies employing high heating rates yield data that support diffusion-controlled tar formation (6,8).

Mass spectrometric studies of coal extracts and vitrains have yielded some interesting results (9). Between 90°C and 150°C, only the lowest-rank coals had appreciable spectra. The products that were evolved from the low-rank coals were largely aliphatic in nature. Aliphatic hydrocarbons of a wide molecular weight range were evolved between 200°C and 370°C. The yield of aliphatic compounds was found, however, to decrease rapidly with increasing coal rank. Alkylbenzenes and alkylnaphthalenes were also found among the products in the 200–370°C temperature range. Because of the long duration of the observed spectra, it was postulated that the simple aromatic compounds were diffusing from the micropores of the coal. An increase in the quantity of material in the 450–650 molecular weight range was observed at still higher temperatures. Alkylphenols and naphthols were also identified among the pyrolysis products. It was postulated that the phenolic compounds originated from the thermal cleavage of benzylic ether–type bonds. This would indicate that the onset of structural decomposition of the coal matrix begins at temperatures around 400°C and above.

Pyrolysis of coal gives us information about the changes in organic composition of coal as a function of rank. There is also good evidence that coal contains both a covalently bonded polymer matrix and a quantity of organic material that is not covalently bonded to the matrix. Both of these organic materials are important in the overall chemistry and properties of coal.

3.2 SOLVENT EXTRACTION OF COAL

Solvent extraction of coal is a second technique that has been widely employed to study the structure of coal. The use of solvent extraction as a tool for the study of coal dates from the observation by P. P. Bedson that bituminous coals are relatively soluble in hot pyridine (10). I. G. C. Dryden has pointed out that the relative efficiency of coal solvents are related to the availability of an unshared electron pair on an oxygen or nitrogen atom of the solvent molecule (11). Dryden designated such coal solvents "specific" coal solvents. Solvents lacking this electron pair were designated "nonspecific" coal solvents. The solubility of a given coal depends not only on the type of solvent employed but also on rank, petrographic composition, extraction temperature, and the extraction procedure employed. With the specific coal solvent pyridine, coal solubility appears to pass through a rather broad maximum around 84–88% carbon and then rather rapidly fall to almost zero at approximately 92% carbon (12). Nonspecific coal solvents usually follow the same general trend but dissolve much less coal (typically 1–10% by weight).

Before continuing the discussion of the solvent extraction of coal and some of the techniques employed, a number of terms commonly used to classify the products obtained from coal extraction and liquefaction need to be defined. Historically, the most common classification scheme is based on the solubility behavior of the coal-derived products. The major classifications of the fractions include *oils* (which are defined as materials that are soluble in hexane), *asphaltenes* (which are defined as soluble in benzene but insoluble in hexane), and *preasphaltenes* (which are defined as soluble in pyridine but insoluble in benzene). Admittedly, this is a rather crude classification scheme with a good deal of overlap, but it is widely used. The solvents employed are also somewhat variable. Frequently pentane or heptane will be used in place of hexane. Toluene is now frequently used in place of benzene, and other "specific" solvents are sometimes used in place of pyridine.

We will discuss five basic types of coal extraction procedures. The first is generally called *nonspecific extraction.* This procedure is carried out with solvents such as benzene and chloroform at temperatures usually below 100° C. The yield of extractable products is normally in the range from 1% to 10% of the dry, ash-free weight of the coal. The second procedure is often called *specific extraction.* This technique employs specific coal solvents, such as pyridine, that are nucleophiles and bases, and have electron-donor capacity and hydrogen-bond-forming capacity. The extraction temperature is generally below 200° C, and the yield of extractable material is typically between 10% and 30% of the dry, ash-free weight of the coal. The third procedure is called *extractive pyrolysis.* This extraction may use either specific or nonspecific coal solvents at temperatures above the decomposition point of the coal matrix (450° C). This procedure is actually a pyrolysis reaction carried out in the presence of solvent and can therefore lead to reasonably high yields of extraction products. A typical yield may be around 30%, but the actual yield depends on the nature of the solvent employed, the rank of the coal being extracted, and the extraction temperature. The fourth and probably most widely studied procedure may be called *reductive extraction.* This is carried out with solvents that are good hydrogen donors. Tetrahydronaphthalene (tetralin) is an example of such a solvent and has been widely used as a hydrogen donor solvent:

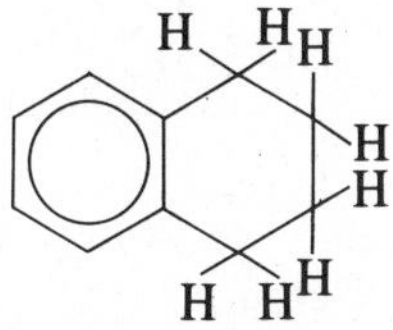

H H H H H H H H

Tetrahydronaphthalene
(tetralin)

In addition, this extraction may also be carried out in a hydrogen atmosphere in the presence of a hydrogenation catalyst. In this pyrolytic technique, the extraction temperature is typically around 450° C or above. The yield of extractable material can be in excess of 90% of the dry, ash-free weight of the coal. Finally, the recent use of *supercritical coal extraction* has proven very useful in the study of coal structure. The use of solvents at temperatures above their critical temperature reportedly results in unusual solvent properties (13,14). Critical temperature is the temperature above which a given gas cannot be liquefied regardless of the pressure applied. Even nonspecific coal solvents, such as hydrocarbons, can give extraction yields of around 20% of the dry, ash-free weight of the coal when used in supercritical extractions.

Nonspecific extraction of coals with solvents such as benzene and chloroform give the lowest yield of extraction products. Since the extraction temperature is too low (100°C or less) to cause much, if any, decomposition of the coal matrix, these extraction products are most likely those substances that are contained within the pore structure and channels of the coal particle. Normally the products obtained in the benzene extraction of a coal or maceral include alkanes, alkenes, cycloalkanes, hydroaromatics, alkylbenzenes, and alkylnaphthalenes. The actual products obtained depends on the rank of the coal or the maceral. Rather large polynuclear aromatic systems, such as derivatives of picene and chrysene, have, however, been isolated in very low yields from the benzene extract of a bituminous coal:

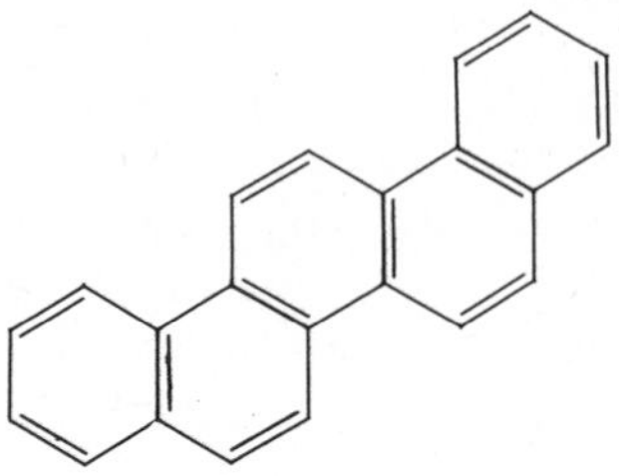

Picene

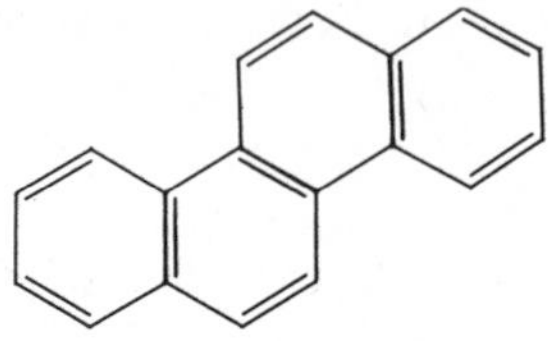

Chrysene

Chloroform extracts of vitrinite from a lower-rank coal have yielded normal or linear alkanes, methyl-substituted linear alkanes, isoprenoid hydrocarbons, alkylbenzenes, and various derivatives of naphthalene and higher aromatics. Isoprenoid hydrocarbons are formally based on isoprene ($CH_2{=}C(CH_3)CH{=}CH_2$) units. The isoprene unit is a common building block in nature. Nearly all terpenes (found in plants) have carbon skeletons made up of isoprene units joined in a regular manner.

Nonspecific coal solvents may be much more effective when used under supercritical conditions. One technique, for example, uses toluene at approximately 350°C (the critical temperature for toluene is 318°C) to extract approximately 17% of the dry, ash-free weight of a bituminous coal (15). This extraction takes place at temperatures below those at which extensive thermal decomposition of the coal matrix is known to occur. In addition, the procedure employs a nonspecific coal solvent. The toluene extract was found to contain approximately 0.4% alkanes plus a smaller amount of cycloalkanes. The majority of the alkanes were straight-chain alkanes showing a predominance of those with odd carbon numbers over those with even carbon numbers. The predominance of compounds with an odd number of carbon atoms is a key feature of residues of biological origin. Also found in the toluene extract were 2-methylalkanes, 3-methylalkanes, and cycloalkanes. In addition, the isoprenoidal hydrocarbons phytane, pristane, norpristane, and farnesane were among those identified.

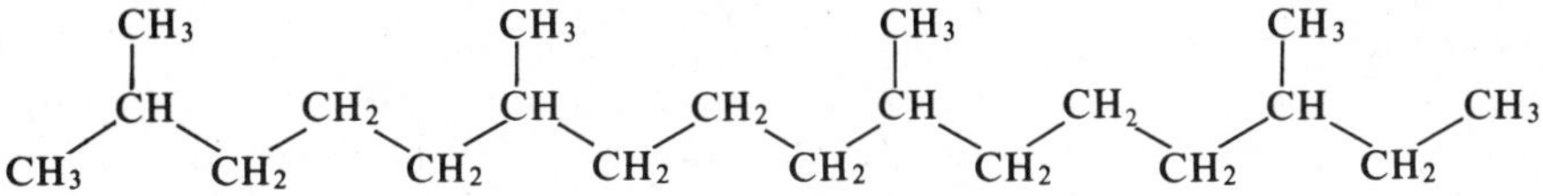

2,6,10,14-Tetramethylhexadecane (phytane)

The bulk of the remainder of the coal extract (16% of the dry, ash-free weight of the coal) was found to be aromatic in nature. Derivatives of benzene, naphthalene, biphenyl, phenanthrene, pyrene, benzanthracene, and fluoranthene were among those identified in the aromatic portion of the coal extract.

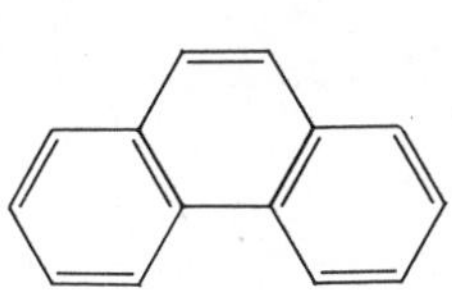

Phenanthrene

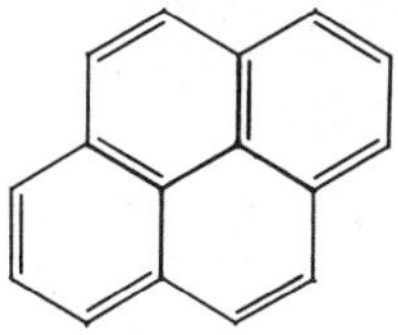

Pyrene

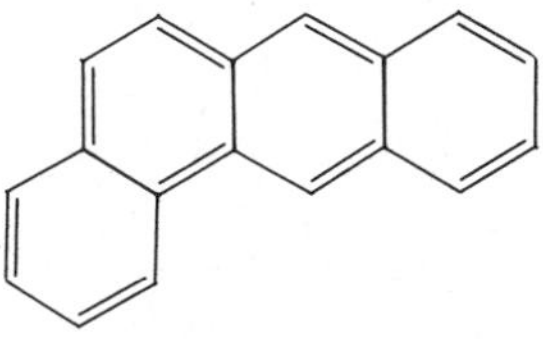

1,2-Benzanthracene

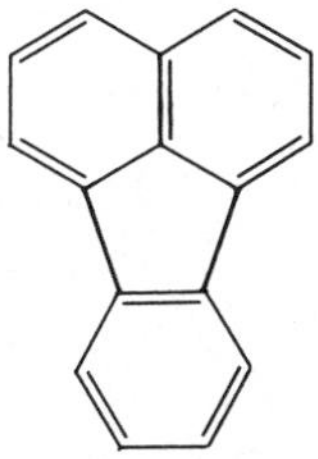

Fluoranthene

The longest alkyl substituent found on the aromatic system was butyl, with methyl being by far the most common. Some of the aliphatic material in the aromatic fraction was also present in the form of hydroaromatic structures within a ring system.

The chemical nature of the coal extract from nonspecific extraction (normal and supercritical) is consistent with the view that these compounds separated from the parent coal with little, if any, structural alteration. The presence of the complex hydroaromatic structures, isoprenoid, and straight-chain alkanes and the absence of straight-chain alkenes strongly suggests that the bulk of the extractable material actually arose from the plant coalification process and not from pyrolytic decomposition of the coal during extraction. The supercritical gas probably allows better penetration of the pore structure of the coal and thus the extraction of materials not normally recovered in a typical nonspecific extraction. The extraction temperature of the supercritical extraction is also significantly higher than in a typical nonspecific extraction. This experiment also illustrates the common belief that a substantial amount of the organic material in coal (10–25%) is not covalently or ionicly bonded to the coal matrix. This material may be trapped within the extensive capillary system or complexed with functional groups on the coal matrix.

Pyridine is one of the more extensively used solvents for the specific extraction of coal. For example, pyridine will extract over 20% of a bituminous coal containing 84–88% carbon. Lignitic coals and higher-rank coals generally yield less pyridine extract. Pyridine appears to have a maximum extraction yield with coal of approximately 88% carbon content. With increasing carbon content, the yield falls off very rapidly to essentially zero at approximately 92% carbon. It has been shown that the infrared spectra of pyridine extracts of coal closely resemble the spectra of the parent coals (16). This finding indicates that the extracted material retains the structural characteristics of the parent coal. Analysis of the pyridine extracts should, therefore, provide very useful information about the structure of the coal itself.

It has been proposed that the enhanced solubility of coal in pyridine is at least partially due to the swelling of the coal particles that occurs on treatment with this and similar solvents (17). The enlarged pore structure should then allow for a more efficient removal of the organic material contained within the capillary system. Coal also imbibes quantities of pyridine. This process brings the solvent into the solid, disrupting hydrogen bonds and increasing the quantity of extractable material. In addition, the polar and nucleophilic character of pyridine allows solubilization of materials not normally soluble in low-polarity solvents such as benzene and toluene. However, a portion of the coal taken up in pyridine is not actually dissolved but is present in the form of a colloidal suspension. While nonspecific solvent extraction of coal yields largely nonpolar materials classified as oils, the pyridine extract of a bituminous coal contains materials classified as oils, asphaltenes, and preasphaltenes. Some of these larger species may not be actually dissolved in the pyridine but may be present as a dispersion in the solvent.

Studies of pyridine extracts of coals of various ranks have yielded valuable information about the trends in the chemical makeup of the parent coal. Aromaticity (f_a) is that fraction of the total carbon content that is part of an aromatic system. As the carbon content of the parent coal increases from about 61% to 86%, the aromaticity (f_a) of the pyridine extract increases from about 0.40 to 0.70. For still higher-rank coals, the aromaticity of the pyridine extract increases rapidly to a value of approximately 0.88.

The number of condensed aromatic rings in a given structural unit may be represented by the symbol R_n. This number is also found to be rank dependent. This number has been found to be approximately 1 for peat and lignite, roughly 2–4 for the pyridine extracts from coal with 73–86% carbon, and about 5–7 for pyridine extracts from coal with 87–91% carbon content.

The degree of alkyl substitution of the aromatic ring system has been observed to decrease with an increase in coal rank. Very notable in the ^{13}C NMR spectrum of pyridine coal extracts is the Ar—CH_2—CH_2—Ar linkage (Ar represents an aromatic ring system), such as is found in 9,10-dihydrophenanthrene:

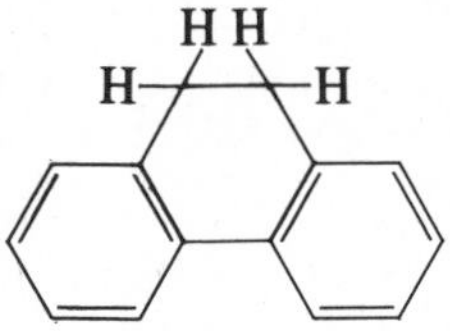

9,10-Dihydrophenanthrene

This is, therefore, very likely an important structural feature in bituminous coals. NMR spectroscopy and infrared spectroscopy also support the idea that low-rank coals are made up of smaller-bridged aromatic units, whereas in the higher-rank coals, a pericondensed (fused-ring) aromatic system would appear predominant.

Extractive pyrolysis and reductive extraction are both carried out at temperatures around the decomposition point of the coal matrix (around 400° C). These two techniques will be discussed together. The use of hydrogen-donating solvents and high reaction temperatures (400° C and above) can dissolve in excess of 90% of the coal on a dry, ash-free basis. It is obvious that under these conditions the polymeric coal matrix is undergoing decomposition to smaller fragments. The aliphatic and aromatic hydrocarbons, as well as the smaller asphaltenes within the pore structure of the coal, are also dissolved.

The solvent employed in the extraction is imbibed by the coal and enters the pore system. The solvent is, therefore, brought into intimate contact with the coal matrix. Various C—C, C—O, and C—S bonds can undergo thermolysis at temperatures above 400° C (Table 3.2). Scission of one or more of these bonds followed by hydrogen transfer from the solvent results in the formation of smaller and more soluble units.

Tetralin is an example of a very good hydrogen-donating solvent that has been widely employed in the reductive extraction of coal. A subbituminous coal, for example, was converted into approximately 80% tetrahydrofuran-soluble products within 2 h by reaction with tetralin at 427° C (17). Molecular weight studies (using vapor pressure osmometry) indicate a drop in the number-average molecular weight from approximately 1,200 at 10 min reaction time to approximately 500 at 30 min and longer reaction time. These data suggest an initial rapid cleavage of relatively few "weak" bonds to form preashaltenes. This cleavage is followed by a further conversion of the preasphaltenes at a slower rate to the lower molecular weight asphaltenes and oils.

It is also known that the lower-rank coals very rapidly lose about 20% of their oxygen content on hydrogenation or reductive extraction. Most of this oxygen is likely lost through some dehydration process. A significant portion of the oxygen present in these coals is believed to be in the form of ether groups (R—O—R) bonded to an aromatic ring. It is very probable, therefore, that ether linkages are very important structural units in the lower-rank coals. Many of the covalent bonds undergoing thermolysis at 400°C and above are likely benzylic ethers (Ar—CH_2—O—Ar, Ar—CH_2—O—CH_2—Ar) and activated aliphatic bridges between aromatic rings. The hydroxyl group (—OH), for example, is very effective in accelerating the thermal cleavage of the methylene bridge (—CH_2—) of diphenylmethane (C_6H_5—CH_2—C_6H_5) (4).

In summary, there are three basic categories of compounds isolated in the solvent extraction of coal: oils, asphaltenes, and preasphaltenes. The overall extraction yield and the amount of material found in each category depends on the nature of the coal, the solvent used, the extraction temperature, and the extraction technique employed. The oils and asphaltenes are substances of relatively low molecular weight (generally 500 or less), while the presphaltenes tend to have a significantly higher molecular weight ($>$ 1,000). The oils consist largely of hydrocarbon materials. These consist of a rather wide range of aliphatic and aromatic compounds. The lower-rank coals tend to contain a greater percentage of aliphatic compounds than do the higher-rank coals. In addition to the hydrocarbons, there are also smaller amounts of a variety of organic compounds, including aromatic compounds containing ether groups and oxygen, sulfur, and nitrogen heterocycles.

Extraction studies have also shown a preferred sequence for the evolution of the hydrocarbons from coal. The cycloalkanes and branched alkanes are initially favored over the simpler straight-chain alkanes. Likewise, the more complex aromatic hydrocarbons tend to initially predominate over the less complex ones. This selectivity can be explained on the basis of a "molecular sieve effect" (18). That is, the hydrocarbons with larger critical dimensions show a greater ease of evolution from the capillary system of coal.

The asphaltenes have a relatively high oxygen content, especially in the form of phenolic hydroxyl groups (the —OH group attached directly to an aromatic ring). Heterocyclic and ether linkages are also present. As we shall see later, the asphaltenes often exist as acid-base complexes and can be resolved into acidic and basic components. Small amounts of organic nitrogen and sulfur-containing compounds are also found among the asphaltenes. The majority of the compounds in the asphaltene class are aromatic and hydroaromatic in nature. These substances are more frequently observed in high yields on extraction of coal, especially with nucleophilic solvents, than on low-temperature pyrolysis. These materials are more polar and less volatile than are the oils. The asphaltenes may be found both in the capillary system and bound to the polymeric coal matrix via hydrogen bonding and acid-base interactions. Asphaltenes of this kind are also frequently observed at the onset of coal decomposition, and some may be derived by the thermal rupture of bonds of the benzylic ether type (Ar—CH_2—O—Ar).

The preasphaltenes are distinguished from the oils and asphaltenes by their higher molecular weight (generally > 1,000) and their higher heteroatom (oxygen, nitrogen, and sulfur) content. There are, for example, more hydroxyl (—OH) groups per preasphaltene unit than are commonly found in oils or asphaltenes. Accordingly, the preasphaltenes have also been called asphatols (19). These substances are believed to be the materials initially formed on the thermal decomposition of coal. It is also possible that many of the —OH groups arise from the thermolysis of ether linkages originally present in the parent coal.

3.3 HYDROGENATION OF COAL

The results of hydrogenation studies of coal closely parallel those obtained from studies of the reductive extraction of coal. Since the processes are basically the same (the addition of hydrogen to coal), this fact is not surprising. The hydrogenation of coal is thought to proceed through the same basic sequence of reactions irrespective of the presence of catalyst or hydrogen-donor solvent. The most widely accepted mechanism is again based on the thermal generation of free radicals by the scission of certain covalent bonds of the coal matrix (Fig. 3.1). At temperatures of 400°C and above, the reaction appears to be largely independent of coal particle size. The hydrogenation reaction also appears to be chemically controlled, with the pyrolytic bond cleavage being the principal driving force. The subsequent progress of the reaction depends on the interaction of the thermally generated free radicals with hydrogen. In the absence of a hydrogen source, the radicals may disproportionate to form smaller volatile molecules or may undergo radical coupling to yield char or coke rather than liquid or gaseous products. The effect of added hydrogen on the overall yield of volatile products can be clearly seen in Table 3.1.

In addition to the thermolysis of bonds linking structural units, dealkylation, hydrodesulfurization, and hydrodeoxygenation reactions have also been observed. It should be noted that, while the oxygen and sulfur content of coal appear to decrease on hydrogenation, the nitrogen-containing structures appear to be largely inert toward hydrogenolysis. This leads to problems in coal liquefaction or, more correctly, in the utilization of the coal-derived liquids. Hydrogenation of coal and the nature of the coal-derived liquids are covered in more detail in Chapter 4.

3.4 ALKYLATION OF COAL

A number of techniques have been developed to study coal structure and to depolymerize the coal matrix. These have been reviewed by J. W. Larsen and E. W. Kuemmerle (20) and by R. M. Davidson (21). The depolymerization of coal with phenol-BF_3 has been a widely used technique (22–25). The function of the BF_3 is that of an acid catalyst. The function of phenol in these experiments has been attributed to several factors. First, phenol can be alkylated by the coal fragments produced by the acid-catalyzed depolymerization of the coal matrix:

$$\text{C}_6\text{H}_5\text{OH} + \text{R—Y} \xrightarrow{BF_3} \text{R—C}_6\text{H}_4\text{OH} + \text{C}_6\text{H}_5\text{O—R} + \text{H—Y}$$

Alkylation is the addition or transfer of an alkyl group to an atom or molecule. Second, phenol can penetrate the micropores of the coal particles for intimate contact with the coal matrix and also can promote swelling and subsequent extraction. Finally, it has been shown that activated aromatic-methylene bonds and ether bonds are cleaved under reaction conditions in the presence of phenol (26). Activated aromatic-mythylene bonds are those that contain an electron-donating group, such as hydroxyl, attached to the aromatic ring. The data presented thus far suggest that ether linkages and activated aromatic-methylene bonds are significant bridging groups in lower-rank coals.

More recent work by J. W. Larsen (27) and H. P. Hombach (28) has shown that the amount of colloidal material present in the coal extracts from these depolymerization experiments is large. Larsen (26,27) has attributed the failure of the phenol-acid depolymerization of coal to the fact that bonds cleaved under the reaction conditions employed are not plentiful and to the fact that acid-catalyzed repolymerization reactions likely take place.

Friedel-Crafts alkylation and transalkylation have also been used to study coal structure. Friedel-Crafts alkylation is the substitution of an alkyl group (carbon chain) on an aromatic ring in the presence of an acid catalyst:

$$\text{C}_6\text{H}_6 + (CH_3)_2CHCl \xrightarrow{AlCl_3} \text{C}_6\text{H}_5\text{—}CH(CH_3)_2 + HCl$$

For example, it has been shown that Illinois Number 6, a bituminous coal, was alkylated by 2-chloropropane in the presence of aluminum chloride under very mild conditions (29). A significantly enhanced coal solubility resulted from the alkylation; the results of this study are shown in Table 3.3.

Illinois Number 6 and Pittsburgh Seam, also a bituminous coal, were alkylated successfully using gaseous HF as the catalyst and a variety of alcohols as alkylating agents (30). The enhanced solubility of the alkylated coal is believed to occur via cleavage of ether linkages in the coal during the alkylation process (31). The results of the HF-catalyzed alkylation reactions can be seen in Table 3.4. The increase in solubility on treatment with HF alone has been attributed to acid-catalyzed cleavage of certain ether linkages within the coal matrix. The principal disadvantage associated with the use of the HF-alcohol system for the alkylation of coal is the tendency of the alcohol to undergo condensation and polymerization. In light of the findings of Larsen (27) and Hombach (28), one might question the reported 99% solubility of the alkylated Illinois Number 6 coal in pyridine.

K. Ouchi and coworkers (32) have also reported successful alkylation of a vitrinite concentrate of Taiheyio coal (Japan) using a $ZnCl_2$-alcohol system. They reported branched-chain alcohols to be superior to straight-chain alcohols and that the solubility of the alkylated coal increased with increasing chain length of the alcohol employed.

Transalkylation has recently been used as a tool for the study of alkyl substituents on the aromatic rings of coal (33). Transalkylation involves the transfer of an alkyl group (carbon chain) form one aromatic ring (in coal) to another (the aromatic acceptor employed in the study). Wyodak, a subbituminous coal, was treated with toluene (aromatic acceptor molecule) in the presence of trifluoromethanesulfonic acid (CF_3SO_3H) as the acid catalyst. The toluene solution was collected, dried, and subjected to gas chromatographic analysis. A complex mixture of substituted toluene derivatives was observed. Normal and branched alkyl chains of up to six carbons were identified as present on the aromatic ring. No chain longer than six carbons was observed,

TABLE 3.3 Friedel-Crafts Alkylation of Coal with 2-Chloropropane-$AlCl_3$

	Percent Solubility		
Coal	Benzene	Pyridine	H/C
Raw coal[a]	2	11	0.82
Alkylated coal (5°C)	11	30	0.94
Alkylated coal (150°C)	15	39	0.98

Source: Reference 29; reprinted with permission from R.H. Schlosberg et al., "Friedel-Crafts Isopropylation of a Bituminous Coal under Remarkably Mild Conditions," *J. Am. Chem. Soc., 100,* 4188, 1978, American Chemical Society.

[a] 77.66% C, 5.36% H, 4.75% S, 1.36% N, 10.88% O.

TABLE 3.4 Alkylation of Coal with Alcohol-HF

Coal	Tetrahydrofuran Solubility (wt%)	Pyridine Solubility (wt%)
Pittsburgh Seam[a]	8	20
Pittsburgh Seam (HF only)	20	30
Pittsburgh Seam (HF-butyl alcohol)	59	83
Illinois Number 6[b]	8	15
Illinois Number 6 (HF only)	10	20
Illinois Number 6 (HF-2-butanol)	61	99

Source: Reference 30; reprinted with permission from R.A. Flores et al., "Hydrogen Fluoride-Catalyzed Alkylations of Pittsburgh Seam (hvb) Coal," *Fuel, 57,* 697, 1978, Butterworth & Co., Publishers, Ltd.

[a] 78.5% C, 5.1% H, 1.5% N, 2.1% S, 11.1% O, 1.7% ash.
[b] 73.4% C, 4.9% H, 1.5% N, 3.2% S, 6.6% O, 10.4% ash.

and ethyl (two carbons) was found to be the most abundant. This technique shows promise for the examination of the kinds and quantities of alkyl groups bound to coal. It should also be noted that transalkylation was reported not to increase coal solubility.

Finally, let us consider briefly the reductive alkylation of coal. This process concurrently may add electrons (reduction) to the coal structure and attach carbon chains (alkyl groups). The most popular technique is that developed by H. W. Sternberg (34,35). The treatment of many coals with alkali metals in tetrahydrofuran in the presence of naphthalene ($C_{10}H_8$) yields polyanions or "coal anions," which may then be alkylated by reaction with alkyl halides:

$$\dot{Na} + C_{10}H_8 \longrightarrow Na^+ + C_{10}H_8^{\dot{-}}$$

$$\text{Coal} + nC_{10}H_8^{\dot{-}} \longrightarrow \text{Coal}^{n-} + nC_{10}H_8$$

$$\text{Coal}^{n-} + nR\text{—I} \longrightarrow \text{Coal—(R)}_n + nI^-$$

Alkali metals in liquid ammonia have also been employed in reductive alkylation of coal (36,37).

Naphthalene present in tetrahydrofuran (solvent) functions as an electron transfer agent and makes possible the formation of the coal anion. The naphthalene radical anion ($C_{10}H_8^-$) will also cleave ethers, and the reductive cleavage of methylene (—CH_2—) bridges between aromatic rings has been reported (38). These cleavage reactions may cause the coal matrix to depolymerize and react to form a variety of products:

$$\text{Ar—O—R} + C_{10}H_8^{\dot{-}} \longrightarrow \text{Ar—O}^- + \dot{R} + C_{10}H_8$$

The actual solubility of reductively alkylated coal is reported to depend on the rank of the coal. H. Wachowska studied the reduction of a number of coals using potassium in tetrahydrofuran in the presence of naphthalene (39). The number-average molecular weights for benzene extracts of lower-rank (78.2% carbon) coals ranged roughly from 500 to 800. The higher-rank (87.9% carbon) coals, however, gave benzene extracts with number-average molecular weights in the 1,300–2,000 range. This finding was explained in terms of an increase in the degree of condensation of the aromatic system with increasing rank. If the cleavage of ether and methylene bridges is occurring during the depolymerization of the coal matrix, then the molecular fragments between bridging bonds is significantly larger. Again, it should be pointed out that the presence of colloidal material in the coal extract can yield incorrect molecular weights (too low) as well as abnormally high solubility.

3.5 OXIDATION OF COAL

Coal is an extremely reactive material toward oxidation. Even small amounts of air (oxygen) can significantly alter the physical and chemical properties of a coal (40). When coal is oxidized, complex mixtures of materials called humic acids are formed. These humic acids are readily extracted from oxidized coal with aqueous alkali. When the alkaline solution is acidified, the humic acids precipitate as brown solids. These solids dry to form brittle, coallike solids. The spectral characteristics of the dry humic acids is very similar to those of the parent coal. It would appear, therefore, that the humic acids retain much of the original coal structure. Further oxidation of the humic acid material ultimately yields a mixture of aliphatic compounds and benzenecarboxylic acids. Alkyl-substituted benzene rings and polynuclear aromatic compounds are known to yield benzenecarboxylic acids on oxidation with strong oxidants:

$$C_6H_5CH_3 \xrightarrow{[O]} C_6H_5CO_2H$$

$$C_6H_5CH_2CH_3 \xrightarrow{[O]} C_6H_5CO_2H + CO_2$$

$$C_{10}H_8 \xrightarrow{[O]} C_6H_4(CO_2H)_2 + 2CO_2$$

Oxidation has been one of the more widely used techniques for the degradation of coal into smaller molecules, which can then be identified and interpreted in terms of the overall coal structure. A large number of oxidizing agents have been tried, most of them being very strong oxidants that convert most of the aromatic material present in the coal to a mixture of benzenecarboxylic acids. This kind of oxidation is of limited usefulness but can give worthwhile information about the aromaticity and degree of substitution of the aromatics present in coal. For example, nitric acid (HNO_3) yields mostly benzenecarboxylic acids. The number of carboxylic acid groups per ring correlates well with the degree of substitution or condensation of the aromatic system within the coal. Oxidation with alkaline potassium permanganate ($KMnO_4$) also yields a variety of benzenecarboxylic acids. The presence of relatively large amounts of benzenetricarboxylic, benzenetetracarboxylic, benzenepentacarboxylic, and benzenehexacarboxylic acids indicates that the aromatic rings are rather highly substituted or condensed, especially in the higher-rank coals.

R. Hayatsu and coworkers have extensively studied coal structure using oxidation as a degradation technique (41–43). Aqueous sodium dichromate ($Na_2Cr_2O_7$) is a more selective oxidizing agent and oxidizes alkyl groups and hydroaromatic groups with little degradation of the aromatic ring. Heterocyclic rings have also been detected in the dichromate oxidation products of coal.

Hayatsu converted the carboxylic acids to methyl esters to improve volatility and analyzed the mixture via gas chromatography coupled with time-of-flight mass spectrometry. A variety of aromatic carboxylic acids were identified, including derivatives of benzene, naphthalene, phenanthrene, and heterocyclic aromatic rings containing oxygen, nitrogen, and sulfur.

The observation that no polynuclear aromatic compounds with more than two fused rings were identified among the oxidation products from a lignite is further evidence that few, if any, highly condensed structures are present in the parent lignite. This is consistent with solvent extraction studies, which indicate an R_n of 1–2 for lignites. From the analysis of sodium dichromate oxidation products, it is found that the degree of condensation in the coal matrix increases with increasing rank.

S. K. Chakrabartty and N. Berkowitz have studied coal structure with sodium hypochlorite ($NaOCl$) as the oxidant (44,45). They found no evidence of aromatic compounds other than benzene derivatives in the oxidation products. On the basis of the sodium hypochlorite study, they proposed that coal has a nonaromatic "tricycloalkane" or "polyamantane" structure. This proposal is not in agreement with other data presented on the organic structure of coal. Other workers have suggested that sodium hypochlorite effectively destroys other polynuclear aromatic compounds and gives a misleading picture of coal structure (46–48).

R. Liotta and W. S. Hoff have shown that trifluoroperacetic acid (CF_3CO_3H) converts alkylbenzenes to the corresponding aliphatic carboxylic acid (49):

$$C_6H_5CH_3 \xrightarrow{CF_3CO_2H\text{-}H_2O_2} CH_3CO_2H + 5CO_2$$

No aromatic carboxylic acids were observed. N. C. Deno and coworkers have applied this technique to the study of coal (50–52). Other oxidants typically destroy the aliphatic side chains, leaving the aromatics intact. This method should supplement the previous oxidation studies and allow analysis of the aliphatic component of coal. Oxidation with a mixture of 30% aqueous hydrogen peroxide and trifluoroacetic acid appears to oxidize most of the aromatic material, leaving the bulk of the aliphatic material behind. Among the aliphatic components isolated from the oxidation products of various coals are acetic, malonic, succinic, and glutaric acids. Methanol is also isolated on oxidation of the lowest-rank coals. Deno has also reported isolation of benzenecarboxylic acids from the oxidation products of coal (51).

The acetic acid should arise from methyl groups attached to an aromatic ring. It appears that approximately 1–9% of the carbon present in a bituminous coal is in the form of arylmethyl groups. Malonic acid can be formed via the oxidation of methylene groups joining two aromatic rings. The absence of propionic acid indicates that alkyl chains of two carbons or longer, if present, are present in very small amounts. The formation of glutaric acid results from the presence of either 1,3-diarylpropane bridges or hydroaromatic structures. Likewise, succinic acid should arise from the oxidation of 1,2-diarylethane bridges and hydroaromatic structures. Succinic acid has been reported to be the dominant product from the oxidation of bituminous coals. The methanol isolated appears to arise from methoxy (CH_3O—) groups. While methanol was isolated from the oxidation products of a lignite, it was not detected among the products of the oxidation of bituminous coals. The source of the benzenecarboxylic acids has been attributed to hydroaromatic structures within the coal. Normal alkylaromatics do not yield benzenecarboxylic acids on oxidation with trifluoroperacetic acid.

3.6 HETEROATOMS IN COAL

Before considering the results of the application of instrumental techniques to the structure of whole coal, it would be worthwhile to consider the various forms in which the heteroatoms oxygen, nitrogen, and sulfur are found in coal. There are several reasons for an interest in these kinds of compounds. First, in terms of the overall structure of coal, it is necessary to know the form in which the heteroatoms may be incorporated into the coal matrix. Second, most of the heteroatoms are initially found in the asphaltene and preasphaltene fractions from the degradation of coal. The materials in these fractions are the principal intermediates in the conversion of coal to liquid fuels (liquefaction). They usually have very deleterious effects on the storage of coal-derived liquids, since they tend to cause an increase in the viscosity of the liquid products on storage. Third, the nitrogen- and sulfur-containing compounds are known to poison the catalysts used in the hydrorefining and hydrocracking operations for the upgrading of coal-derived liquids. Fourth, it is also well known that the presence of nitrogen and sulfur in fuels results in high emission levels of the oxides of these substances on combustion.

The oxygen content of coal ranges from a high of 20–30% by weight for a lignite to a low of around 1.5–2.5% by weight for an anthracite. The oxygen content of a given coal is normally determined by difference:

$$\%\ \text{oxygen} = 100 - (\%\ C + \%\ H + \%\ N + \%\ S + \%\ \text{ash})$$

(all calculated on a dry basis)

Oxygen is known to occur in several different forms in coal, including phenolic hydroxyl, carboxylic acid, carbonyl, ether linkages, and heterocyclic oxygen:

OH (on benzene ring)

Phenol

$R{-}CO_2H$

Carboxylic acid

$R{-}C({=}O){-}R'$

Carbonyl

$Ar{-}CH_2{-}O{-}R$

Ether

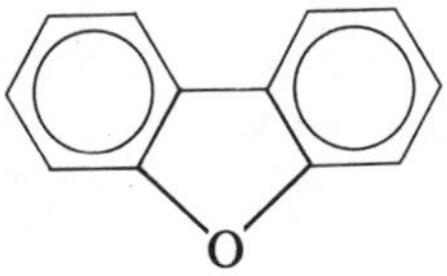

O (dibenzofuran ring)

Heterocyclic oxygen

The presence of phenolic hydroxyl, and especially of carboxylic acid, is highly dependent on the rank or degree of coalification. The content of both these groups decreases rapidly with increasing carbon content (rank). Compounds containing the carboxylic acid group ($—CO_2H$) can be very readily extracted from peats and lignites, but seem to disappear in the subbituminous range. Likewise, the methoxy group ($—OCH_3$) appears to be present in the lignitic coals, but has not been shown to be present in significant amounts in the bituminous and higher-rank coals (50).

The phenolic hydroxyl, ether linkages, and heterocyclic oxygen all appear to be present in bituminous coals and also to a smaller extent in the higher-rank coals. P. H. Given and coworkers have shown that the phenolic hydroxyl content decreases almost linearly with increasing carbon content (53).

The most common form of heterocyclic oxygen is in furan ring systems:

Furan Benzofuran Dibenzofuran

Substituted furan rings have been reported in coal extracts, pyrolysis tars, and oxidative degradation products (43,54). However, to exactly what extent the dibenzofuran system exists in the parent coal and to what extent it is formed by the dehydration of phenols during the degradation of the coal matrix is not clear. The isolation of dibenzofuran from oxidative degradation products and low-temperature coal extracts supports the view that it is present in the parent coal:

OH HO $\xrightarrow{\Delta}$ + H_2O

Nitrogen is typically found in coal in the 0.5–1.5% range by weight. As with oxygen, a number of different types of nitrogen-containing compounds have been isolated from coal-derived liquids (55). A few examples of these include anilines, pyridines, quinolines, isoquinolines, carbazoles, and indoles:

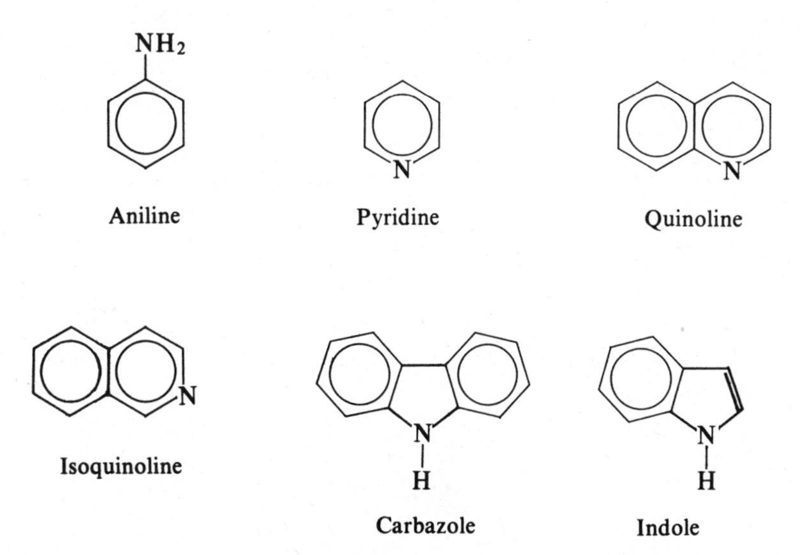

These compounds are frequently substituted with alkyl and aryl groups. F. K. Schweighardt and coworkers analyzed the smaller nitrogen-containing compounds found in the light oil fraction from a hydrogenated bituminous coal (56). A number of methyl-substituted pyridines, anilines, and quinolines were detected. The nature of the pyridine and aniline derivatives in the parent coal is a subject of speculation. If the nitrogen heteroatoms were actually an integral part of the original coal matrix, then hydrogenolysis reactions should ultimately result in the cleavage of a number of carbon-carbon bonds, resulting in the formation of a group of methylpyridines and methylanilines. Figure 3.2 shows one possible way of forming 2,3,6-trimethylpyridine from a hypothetical coal matrix. Aniline can also arise from hydrogenolysis reactions, with the cleavage occurring on the hetero-ring of a fused aromatic ring system in the hypothetical coal matrix. This would then be followed by the cleavage of a carbon-nitrogen bond between nitrogen and an aliphatic carbon. The quinoline ring system could therefore serve as a source of the aniline derivatives. Figure 3.3 shows one way in which an aniline derivative could be formed from the quinoline ring system contained within the coal matrix.

The sulfur content of coal is quite variable, typically somewhere in the range of 0.5–5.0% by weight. This includes both inorganic and organic sulfur. Inorganic sulfur is present mainly in the form of iron pyrite. We shall, however, limit our discussion to the nature of the organic sulfur found in coal. Organic sulfur is commonly defined as all that sulfur which is bonded to the organic molecules in coal.

A number of classes of organic sulfur compounds have been found in the

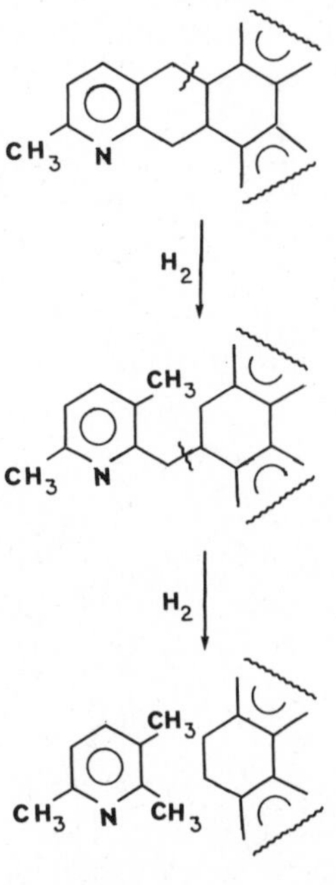

FIGURE 3.2 Possible scheme for the formation of pyridine derivatives from the coal matrix.

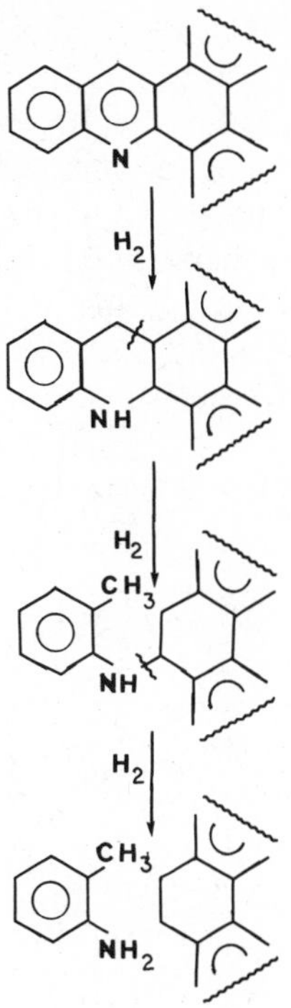

FIGURE 3.3 Possible scheme for the formation of aniline derivatives from the coal matrix.

degradation products of coal, including various derivatives of thiophene, aromatic and aliphatic sulfides, cyclosulfides, thiols, and thiophenols (57,58):

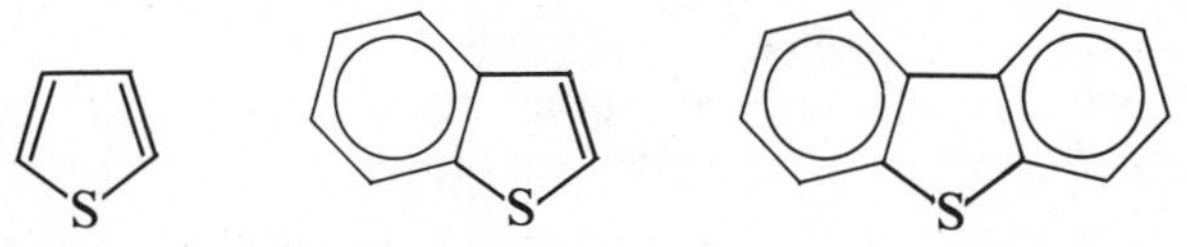

Thiophene 2,3-Benzothiophene Dibenzothiophene

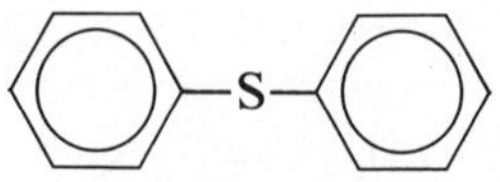

$$R{-}S{-}R'$$

$$R{-}CH_2{-}S{-}H$$

Diphenyl sulfide Alkyl sulfide Thiol

The alkyl, aryl, and alicyclic thiols are present in coal in relatively small amounts and are normally rather easily removed by hydrogenation of the coal-derived liquid product:

$$R{-}S{-}H + H_2 \longrightarrow R{-}H + H_2S$$

The most abundant sulfur functional types found in coal are believed to be derivatives of the thiophene ring system. Other important sulfur types are aryl sulfides, alkyl sulfides, and cyclic sulfides. It should, however, be kept in mind that most of the information on sulfur compounds has been obtained by the analysis of small molecules obtained upon the depolymerization of the coal matrix. The more drastic depolymerization or degradation techniques are very likely to alter the chemical structure of the sulfur compounds originally present in the coal matrix. Hydrogenolysis, for example, removes sulfur from a number of sulfur-containing molecules:

$$\text{S} + 2H_2 \longrightarrow CH_3CH_2CH_2CH_3 + H_2S$$

$$\text{S} + 2H_2 \longrightarrow \text{—}CH_2CH_3 + H_2S$$

$$\text{S} + H_2 \longrightarrow + H_2S$$

3.7 PHYSICAL AND INSTRUMENTAL ANALYSIS OF COAL

All of the techniques reviewed up to this point analyze the smaller fragments obtained via the degradation of the coal structure. A great deal of useful information has been obtained in this manner. Let us now turn our attention to the physical and instrumental techniques that can be used on samples of whole coal. In each case we will be concerned, not with the details of the instrumentation, but with a summary of the results obtained using this technique on coal. While a rather large number of techniques have been employed in the study of the organic structure of coal, we shall concentrate on the information obtained from studies of density, electron spin resonance spectroscopy, x-ray diffraction, and solid state nuclear magnetic resonance spectroscopy of whole coal.

The study of the density of coals and macerals of different rank provides an indication as to the nature of the coal matrix. It has been known for some time that after approximately the 85% carbon level, the density of coal increases continuously until, upon extrapolation to the 100% carbon level, it intersects the ordinate at 2.25 g/cm^3 (59). This is shown in Figure 3.4. The

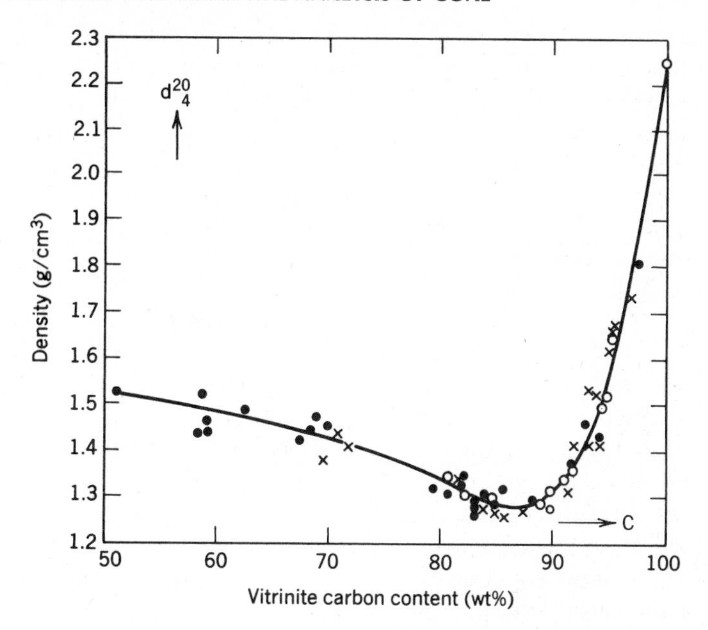

FIGURE 3.4 Vitrinite density as a function of rank. (*Source:* Reprinted with permission from D.W. Van Krevelen, *Coal,* Elsevier, Amsterdam, 1961, p. 314.)

value 2.25 g/cm^3 corresponds to the density of graphite. It would appear, therefore, that for the higher-rank coals, the matrix becomes more and more graphitelike with increasing carbon content (rank).

Electron spin resonance spectroscopy (ESR) has been applied to the study of the organic structure of coal. When a species containing an unpaired electron (free radical) is placed in a magnetic field, the electron spin generates a magnetic field that may be either aligned with the applied field or opposed to the applied field. Two energy states are thus available to the electron: the state in which the magnetic moments are aligned (lower energy) and the state in which the magnetic moments are opposed (higher energy). Energy must be provided (absorbed) in order to promote the electron from the lower to the higher energy level. This energy is supplied to the molecules in the form of electromagnetic (microwave) radiation. This provides an ESR absorption spectrum that yields information about the number and nature of the free radical species present.

ESR studies have been carried out for a series of coals varying in rank from lignite through anthracite. These studies have shown that there are unpaired electrons (free radicals) in coals of all rank. The concentration of free radicals, the spectral line width, and the g values have been shown to be rank dependent (60). The number of unpaired electrons appears to increase from a low of approximately one for each 23,000 carbon atoms (for a sample of peat) to a high of approximately one for each 1,000 carbons (for a sample of anthracite).

This observation is consistent with the view that free radicals in an extended aromatic system are more stable (and thus more abundant) than are those in a less highly condensed system. Lignite (and peat) are believed to have few aromatic systems with more than two condensed rings. Anthracite, on the other hand, is believed to contain more highly condensed aromatic systems (five to seven rings).

The variation in spectral line width observed for the series of coals is thought to be due to spin-spin interactions between the electrons of the free radical and the hydrogens on the carbon of the coal matrix. The ESR line widths do correlate reasonably well with the hydrogen content of the coal sample.

The position of the ESR absorption is indicated as a *g* value. The position of the absorption is a function of the chemical environment of the unpaired electron. The *g* values observed for samples of peat and lignite strongly suggest that at least a portion of the electron spin is delocalized on atoms other than carbon. The *g* values for the higher-rank coals are consistent with those observed for predominantly hydrocarbon free radicals. The *g* values for lignite, however, suggest the involvement of phenoxy-type free radicals. The observed values correlate well with the oxygen content of the coal sample. As the oxygen content decreases with increasing rank, the free radicals become more and more hydrocarbonlike.

X-ray diffraction is a technique used to determine the three-dimensional arrangement of atoms within a structure. The x-rays are scattered by all of the atoms within a structure, and the radiation scattered by each atom interferes with the radiation scattered by all other atoms. This creates an interference pattern that may be interpreted in terms of structure.

X-ray diffraction studies have been carried out on a series of coals of different rank. Lignite shows interference patterns that can be rationalized in terms of a poorly developed system of layers and chains within the coal structure (61). The bituminous coals appear to have structural units that consist of small condensed aromatic ring systems linked to each other by C—C, C—O, and similar linkages to form larger buckled imperfect layers (62). Above the 90% carbon level, the condensed aromatic layers appear to increase rapidly in size with increasing rank. In the lower-rank anthracites, the aromatic layers are roughly 10 Angstrom units (Å) in size (1 Å = 10^{-8} cm). The further increase in rank causes a coalescence of these layers. Higher-rank anthracites, therefore, have layer sizes in the order of 25–40 Å thick and show the development of the three-dimensional reflections that are characteristic of graphite. This indicates a further ordering of layers of molecules of the adjacent layers. In meta-anthracite, the layers are greater than 1,000 Å thick, and the structure of the coal matrix is basically a defective graphite lattice (63).

The data from x-ray diffraction studies indicate that the evolution of coal proceeds from a relatively open structure for the lowest-rank coals to a relatively random structure for the bituminous coals and a well-developed, layered structure rather closely resembling that of graphite for anthracite.

Chemically, the increase in rank is accompanied by a conversion of alicyclic structures to aromatic layers.

Certain atomic nuclei (e.g., ^{1}H and ^{13}C) possess nuclear spin. When such nuclei are placed in an intense magnetic field, the nuclear magnetic moment may either align itself with the applied field (lower energy) or opposed to the applied field (higher energy). Energy must be provided (absorbed) in order to observe transitions between nuclear spin states. This energy is supplied to the molecules in the form of electromagnetic radiation in the radio frequency range. Ths absorption of energy as a function of magnetic field strength provides a nuclear magnetic resonance spectrum. NMR spectroscopy is one of the most powerful tools available to the organic chemist for the study of molecular structure.

A number of coals of different rank have been studied by NMR spectroscopy. Special techniques must be employed in order to obtain useful spectra from whole coal. The technique of ^{1}H-^{13}C cross-polarization is employed to overcome the proton-carbon interactions, and "magic angle" spinning is employed to overcome the chemical-shift anisotropy due to random orientation of the molecules in the coal sample.

The aliphatic region of a ^{13}C NMR spectrum for a high-volatile bituminous coal shows four major resonance bands (64). These bands have been assigned to methyl groups attached to aliphatic moieties (at approximately 15 ppm), methyl groups attached to aromatic moieties (at approximately 20 ppm), cyclic methylenes and certain hydroaromatic carbons (at approximately 29 ppm), and carbons at or adjacent to a highly branched center (between 39 and 44 ppm).

The aliphatic region of the ^{13}C NMR spectrum for a sample of lignite shows the same basic features as does high-volatile bituminous coal with two notable exceptions (64). The resonance band around 42 ppm corresponding to highly branched centers is missing, and the resonance band at 29 ppm corresponding to cyclic methylene and hydroaromatic structures is observed to be more prominent.

Aromaticities for the carbon atoms of coal have been calculated from ^{13}C NMR studies (65). The fraction of the total carbon content that is aromatic in nature (part of an aromatic ring system) is designated by the symbol f_a. The values for f_a have been found to range from around 0.59 for the lowest-rank coals up to around 0.95 for a high-rank anthracite.

The results of the ^{13}C NMR studies can best be rationalized in terms of a progressive increase in both the size and the number of condensed aromatic ring structures present in coal as the degree of coalification increases. This increase occurs at the expense of the nonaromatic material present in the coal (Fig. 3.5). In coals of very high-rank, the aromatic structures become increasingly more graphitelike (66). Thus, the NMR work also presents very strong evidence for an increase in the aromaticity of coals as the degree of coalification (rank) increases. This is, of course, accompanied by a corresponding loss of aliphatic structure (67).

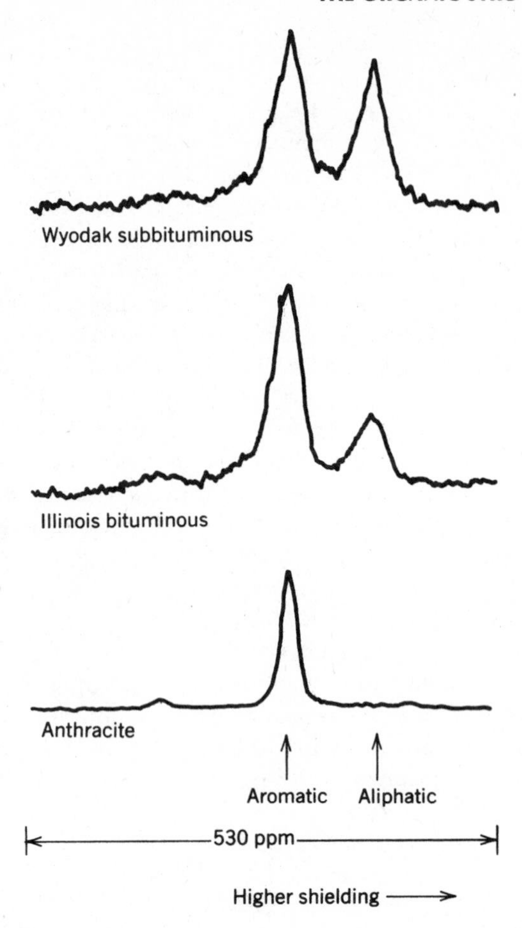

FIGURE 3.5 ^{13}C NMR spectra of whole coal as a function of rank. (*Source:* Reprinted with permission from G.E. Maciel et al., "Characterization of Organic Material in Coal by Proton-Decoupled ^{13}C Nuclear Magnetic Resonance with Magic-Angle Spinning," *Fuel, 58,* 391(1979), Butterworth & Co., Publishers, Ltd.)

3.8 THE ORGANIC MATRIX OF COAL: SOME MODELS

Some examples that can be viewed as models of the types of structural units found in the bituminous coal matrix are shown in Figures 3.6–3.8. These models are not intended to represent "coal molecules," but are simple models of the organic matrix structure.

Probably the most well-known model was developed by P. H. Given (Fig. 3.6) on the basis of x-ray diffraction and other analytical information (68). It is intended to show the three-dimensional aspects (folding or buckling) of the coal structure.

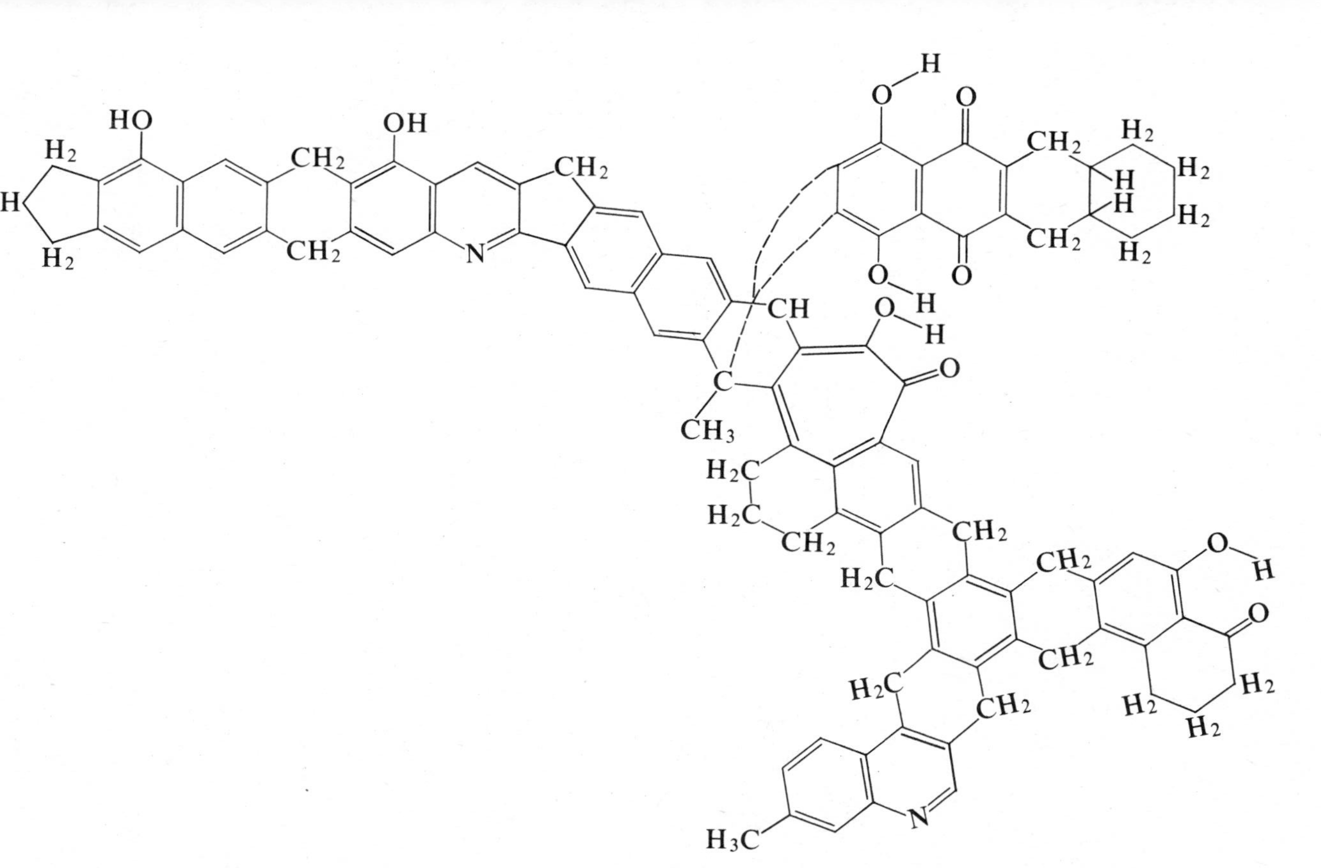

FIGURE 3.6 Given model for bituminous coal. (*Source:* Reprinted with permission from P.H. Given, "A Model Structure for Bituminous Coal," *Fuel, 39,* 147(1960), Butterworth & Co., Publishers, Ltd.)

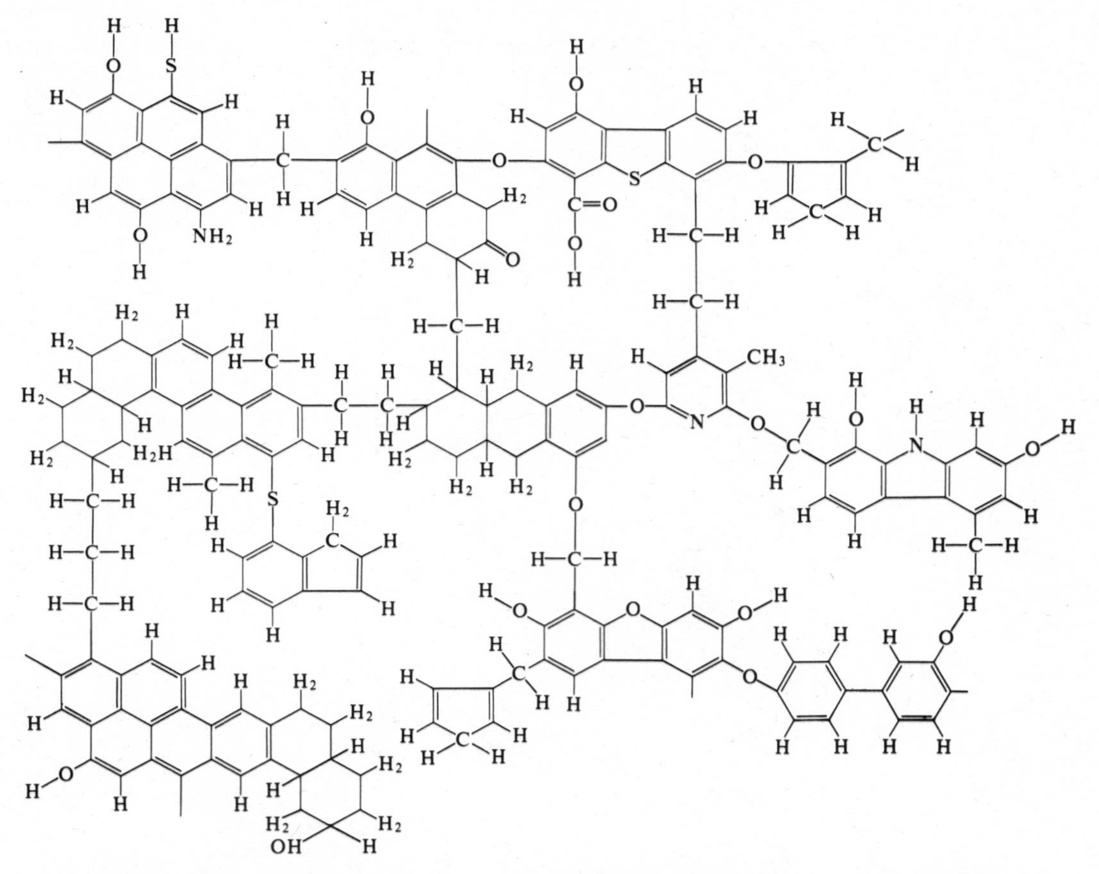

FIGURE 3.7 Wiser model for bituminous coal. (*Source:* Reprinted with permission from W.H. Wiser, "Schematic Representation of Structural Groups and Connecting Bridges in Bituminous Coal," 1978.)

The model developed by W. H. Wiser (Fig. 3.7) illustrates the aromaticity, functionality, and composition of bituminous coal (69). Based on the model, organic chemists can predict chemical behavior. One can, for example, pick out those chemical bonds (Table 3.2) that would undergo thermolysis during a pyrolysis experiment.

The structure proposed by P. R. Solomon (Fig. 3.8) also permits visualization of the processes that occur during pyrolysis (70).

SUMMARY

The data obtained from the chemical studies on coal and the results obtained from physical and instrumental studies on whole coal present an extremely complex yet relatively consistent picture of the organic structure of coal. The organic part of coal most likely consists of a rather complex matrix of large molecules (molecular weights roughly 500–1,000) joined by bridging units such as methylene, ethylene, biphenyl, ether, and sulfide. In addition, smaller amounts (10–25% by weight) of volatile or solvent extractable compounds are noncovalently bonded to the matrix and trapped within the capillary structure of the coal particles. This capillary system arises, at least in part, from the imperfect packing of the large, irregularly shaped molecules of the coal matrix.

The aromaticity of the material found in coal varies with rank from a low around 0.5 for lignite to a high of about 0.95 for anthracite. In addition, it would appear that as the rank of the coal increases, the ordering of the layers of the coal matrix increases from a fairly random orientation in lignite to a rather highly ordered graphitelike three-dimensional lattice for the meta-anthracite coals. Thus, as the degree of coalification increases, the composition of the coal becomes less and less like the plant materials from which it originated and more and more like graphite in structure.

The smaller molecules found within the capillary structure of coal are most likely the result of the normal coalification process. Both lignite and anthracite contain less volatile or extractable organic material than does the bituminous-rank coal. As a matter of fact, anthracite yields little volatile matter other than methane. This is indicative of the highly condensed nature of the anthracite matrix, which has apparently lost its volatile matter in the process of condensation followed by diffusion from the pores through geologic time under the influence of heat and pressure. Lignite shows an abundance of alkanes, terpenoids, and cycloalkanes but relatively few aromatics. It would appear, therefore, that the smaller molecules found in bituminous coals are formed in the normal process of coalification of the lignite. As coalification progresses, the lighter, relatively hydrogen-rich compounds are evolved and then trapped within the capillary system, leaving behind a polymeric matrix that is more aromatic and thus higher in carbon. As coalification progresses, the aromatic character increases further with greater cross-linking and con-

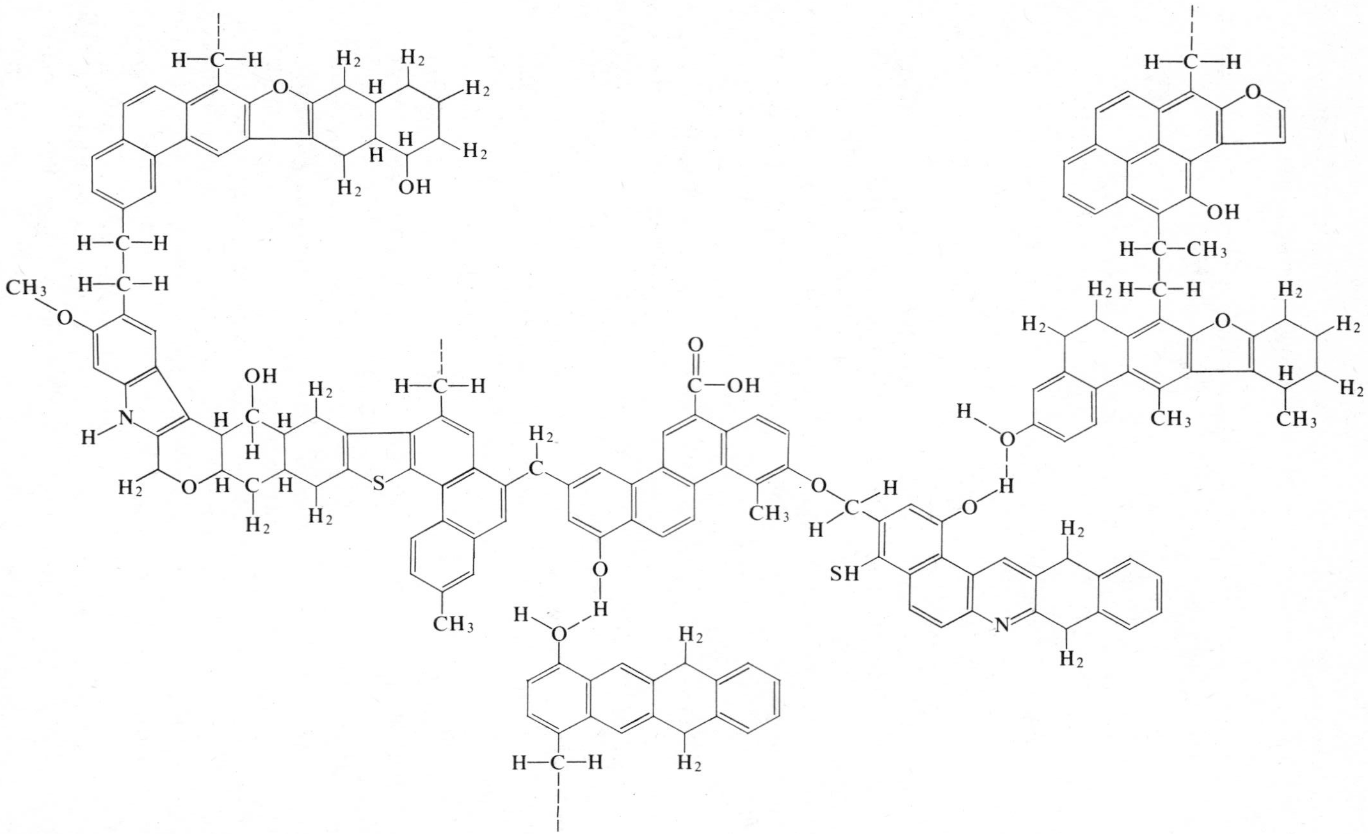

Pyrolysis of this model structure . . .

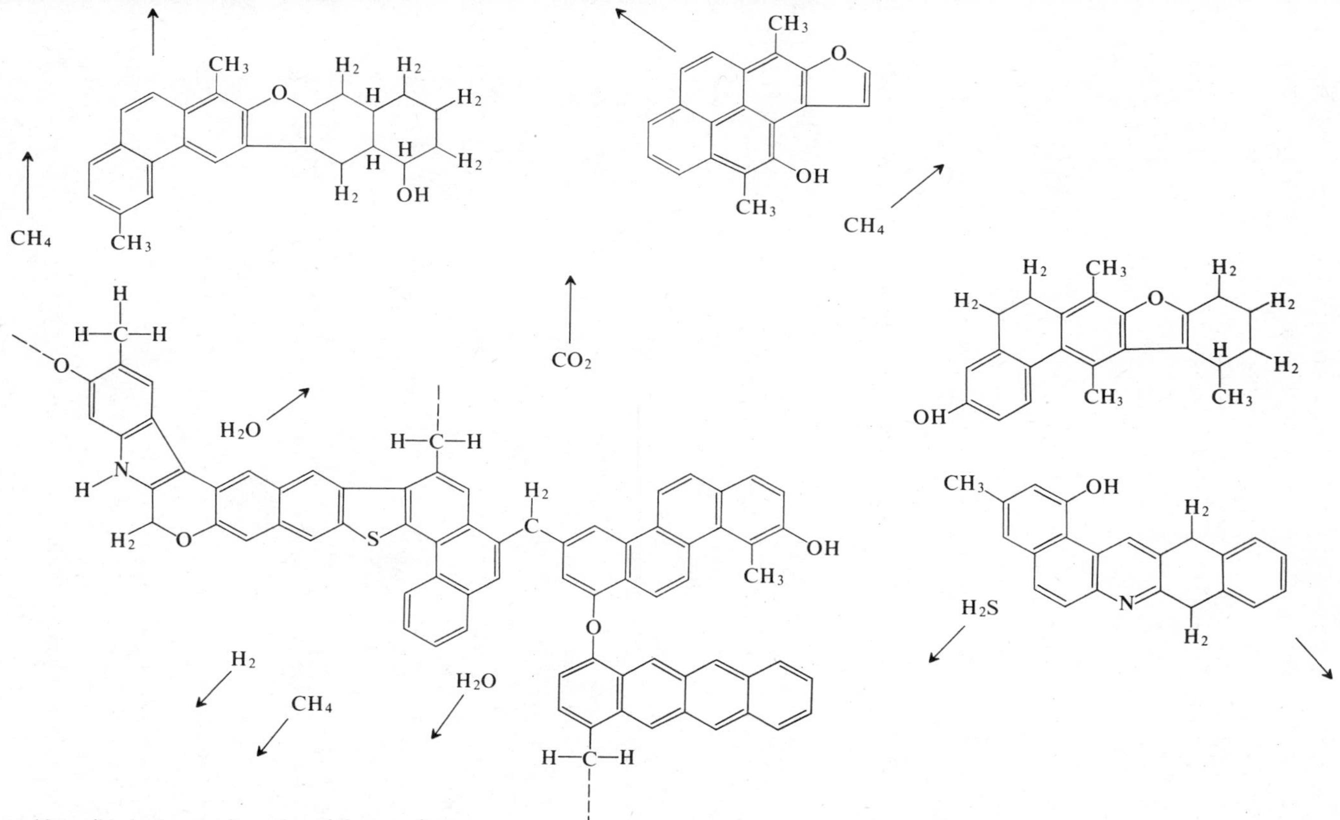

could possibly lead to the formation of these products.

FIGURE 3.8 Solomon model for bituminous coal: application to pyrolysis. (*Source:* Reprinted with permission from P.R. Solomon, "Coal Structural and Thermal Decomposition," *New Approaches in Coal Chemistry,* ACS Symposium Series, *169,* 61, 1981.)

densation until ultimately the coal matrix resembles the crystalline lattice of graphite. The smaller organic molecules generated during coalification gradually become incorporated or diffuse out of the coal over the additional time required for the formation of higher-rank bituminous and anthracite coals. This description provides a reasonable working model for further study of coal. Now that we have some ideas as to the nature of the organic material found in coals of various rank, we can proceed with the study of coal chemistry.

REFERENCES

1. S.R. Gun, J.K. Sama, P.B. Chowdhury, S.K. Mukherjee, and D.K. Mukherjee, *Fuel, 58,* 176 (1979).
2. B.M. Benjamin, F.F. Raaen, P.H. Maupin, L.L. Brown, and C.J. Collins, *Fuel, 57,* 269 (1978).
3. R.E. Miller and S.E. Stein, *Prepr., Div. Fuel Chem., Am. Chem. Soc., 25*(1), 271 (1979).
4. D.F. McMillen, W.C. Ogier, and D.S. Ross, *J. Org. Chem., 46,* 3322 (1981).
5. M. Vahrman and R.H. Watts, *Fuel, 51,* 235 (1972).
6. M.J. McIntosh, *Fuel, 55,* 59 (1976).
7. H. Brusset and B. Le Rat, *Compt. Rend., 238,* 2533 (1954).
8. Vahrman and Watts, p. 130.
9. H.W. Holden and J.C. Robb, *Fuel, 39,* 485 (1960).
10. P.P. Bedson, *J. Soc. Chem. Ind., 21,* 241 (1902).
11. I.G.C. Dryden, *Fuel, 30,* 39 (1951).
12. H.F. Yancey, N.J.F. Johnson, and W.A. Selvig, "Friability, Slacking Characteristics, Low-temperature Carbonization Assay, and Agglutinating Value of Washington and Other Coals," U.S. Bureau of Mines Technical Paper no. 512, 1932, pp. 1–94.
13. J. Whitehead and D. Williams, *J. Inst. Fuel, 48,* 182 (1975).
14. D.F. Williams, *Indian Chem. J., 10,* 20 (1975).
15. K.D. Bartle, T.G. Martin, and D.F. Williams, *Fuel, 54,* 226 (1975).
16. H.H. Lowry, ed., *Chemistry of Coal Utilization,* supplementary vol., Wiley, New York, 1963, p. 75.
17. J.A. Franz, *Fuel, 58,* 405 (1979).
18. M. Vahrman, *Chemistry in Britian, 8,* 16 (1972).
19. D.D. Whitehurst, T.O. Mitchell, and M. Farcasiu, *Coal Liquefaction: The Chemistry and Technology of Thermal Processes,* Academic, New York, 1980, p. 30.
20. J.W. Larsen and E.W. Kuemmerle, *Fuel, 56,* 162 (1976).
21. R.M. Davidson, "Molecular Structure of Coal," in M.L. Gorbaty, J.W. Larsen, and J. Wender, eds., *Coal Science,* vol. 1, Academic, New York, 1982, pp. 83–150.
22. L.A. Heredy and M.B. Neuworth, *Fuel, 41,* 221 (1962).
23. L.A. Heredy, A.E. Kostyo, and M.B. Neuworth, *Fuel, 42,* 182 (1963).
24. L.A. Heredy, A.E. Kostyo, and M.B. Neuworth, *Fuel, 43,* 414 (1964).
25. L.A. Heredy, A.E. Kostyo, and M.B. Neuworth, *Fuel, 44,* 125 (1965).
26. J.W. Larsen and D. Lee, *Fuel, 62,* 463 (1983).
27. Larsen and Lee, p. 918.

28. H.P. Hombach, *Fuel, 61,* 215 (1982).
29. R.H. Schlosberg, M.L. Gorbaty, and T. Aczel, *J. Am. Chem. Soc., 100,* 4188 (1978).
30. R.A. Flores, M.A. Geigel, and F.R. Mayo, *Fuel, 57,* 697 (1978).
31. H.W. Sternberg, *Prepr., Div. Fuel Chem., Am. Chem. Soc., 21*(1), 1 (1976).
32. F. Mondragon, H. Itoh, and K. Ouchi, *Fuel, 61,* 1131 (1982).
33. B.M. Davidson, E.C. Douglas, and D.M. Canonico, *Fuel, 63,* 888 (1984).
34. H.W. Sternberg, C.L. Delle Donne, P. Pantages, E.C. Moroni, and R.E. Markby, *Fuel, 50,* 432 (1971).
35. H.W. Sternberg and C.L. Delle Donne, *Fuel, 53,* 172 (1974).
36. B.S. Ignasiak and M. Gawlak, *Fuel, 56,* 216 (1977).
37. C.I. Handy and L.M. Stock, *Fuel, 61,* 700 (1982).
38. L.B. Alemany, C.I. Handy, and L.M. Stock, *Prepr., Div. Fuel Chem., Am. Chem. Soc., 24*(1), 156 (1979).
39. H. Wachowska, *Fuel, 58,* 99 (1979).
40. F.J. Beafore, K.E. Cawiezel, and C.T. Montgomery, *J. Coal Qual., 3*(4), 17 (1984).
41. R. Hayatsu, R.G. Scott, L.P. Moore, and M.H. Studier, *Nature, 257,* 378 (1975).
42. R. Hayatsu, R.G. Scott, L.P. Moore, and M.H. Studier, *Nature, 261,* 77 (1976).
43. R.C. Duty, R. Hayatsu, R.G. Scott, L.P. Moore, R.E. Winans, and M.H. Studier, *Fuel, 59,* 97 (1980).
44. S.K. Chakrabartty and N. Berkowitz, *Fuel, 53,* 240 (1974).
45. S.K. Chakrabartty and N. Berkowitz, *Fuel, 55,* 362 (1976).
46. G. Ghosh, A. Banerjee, and B.K. Mazumdar, *Fuel, 54,* 294 (1975).
47. T. Aczel, M.L. Gorbaty, P. Maa, and R. Schlosberg, *Fuel, 54,* 295 (1975).
48. F.R. Mayo, *Fuel, 54,* 273 (1975).
49. R. Liotta and W.S. Hoff, *J. Org. Chem., 45,* 2887 (1980).
50. N.C. Deno, B.A. Greigger, and S.G. Stroud, *Fuel, 57,* 455 (1978).
51. N.C. Deno, K.W. Curry, B.A. Greigger, A.D. Jones, W.G. Rakitsky, K.R. Smith, K. Wagner, and R.D. Minard, *Fuel, 59,* 694 (1980).
52. N.C. Deno, K.W. Curry, A.D. Jones, K.R. Keegan, W.G. Rakitsky, C.A. Richter, and R.D. Minard, *Fuel, 60,* 210 (1981).
53. M.B. Abdel-Baset, P.H. Given, and R.F. Yarzab, *Fuel, 57,* 738 (1978).
54. D. Bodzek and A. Marzec, *Fuel, 60,* 47 (1981).
55. J.E. Schiller, *Anal. Chem., 49,* 2292 (1977).
56. F.K. Schweighardt, C.M. White, S. Friedman, and J.L. Schultz, *Prepr., Div. Fuel Chem., Am. Chem. Soc., 22*(5), 124 (1977).
57. A. Attar and W. Corcoran, *Ind. Eng. Chem. Prod. Res. Dev., 16,* 168 (1977).
58. S. Akhtar, A.G. Sharkey, J.L. Shultz, and P.M. Yavorsky, *Prepr., Div. Fuel Chem., Am. Chem. Soc., 19*(1), 207 (1974).
59. D.W. Van Krevelen, *Coal,* Elsevier, Amsterdam, 1961, p. 314.
60. H.L. Retcofsky, J.M. Stark, and R.A. Friedel, *Anal. Chem., 40,* 1699 (1968).
61. R.J. Camier and S.R. Siemon, *Fuel, 57,* 508 (1978).
62. S. Ergun, "X-ray Studies of Coals and Carbonaceous Materials," U.S. Bureau of Mines Bulletin no. 648, 1968, pp. 1–35.
63. S. Ergun, M. Mentser, and H.J. O'Donnell, *Science, 132,* 1314 (1960).
64. K.W. Zilm, R.J. Pugmire, D.M. Grant, R.E. Wood, and W.H. Wiser, *Fuel, 58,* 11 (1978).
65. D.L. VanderHart and H.L. Retcofsky, *Fuel, 55,* 202 (1976).

66. H.L. Retcofsky and R.A. Friedel, *J. Phys. Chem., 77,* 68 (1973).
67. G.E. Maciel, V.J. Bartuska, and F.P. Miknis, *Fuel, 58,* 391 (1979).
68. P.H. Given, *Fuel, 39,* 147 (1960).
69. W.H. Wiser, "Mechanisms of Coal Liquefaction," in *Proceedings of the Department of Energy Project Review Meetings,* June 8, 1978, p. 4.
70. P.R. Solomon, "Coal Structure and Thermal Decomposition," in B.D. Blaustein, B.C. Bockrath, and S. Friedman, Eds., *New Approaches in Coal Chemistry,* ACS Symposium Series, *169,* 61 (1981).

4
COAL CONVERSION

The term *coal conversion* is applied to the process whereby solid coal is modified to yield liquid and gaseous fuels. If the ultimate goal of the process is the production of liquid products from coal, the process is called coal liquefaction. If the ultimate goal is the production of gaseous products, the process is called gasification. Gasification normally implies a quite different engineering approach and different chemistry than does liquefaction.

One of the first questions that should be addressed is the need for coal conversion. Coal conversion is a very complex and extremely expensive undertaking, and a number of technologies are already available to use coal as is for the generation of electricity and of process heat. However, much of the present oil and gas-fired equipment cannot be easily converted to burn coal. This consideration would certainly be important if (some would say when) an oil and gas shortage does occur. In addition, when one considers the fuel needed for transportation and residential heating, the need for a readily available gaseous or liquid fuel is obvious.

The obvious physical disadvantages of coal compared to petroleum are that it is a solid and that it has a high mineral matter content. An objective of coal liquefaction would, then, be to produce a liquid synthetic crude (syncrude) from which a good yield of high-octane gasoline might be obtained. This would require conversion to the liquid state, adjustment of the hydrogen content, reduction of the heteroatom content, and removal of the mineral matter. As we shall see shortly, this would require a high degree of hydrogenation under rather severe conditions. An alternative objective would be the addition of just enough hydrogen to produce a pumpable syncrude with reasonably low oxygen, sulfur, and nitrogen content. Such a coal-derived oil could be used as a boiler fuel for the generation of electricity and similar process heat applications. The availability of such an oil would release an equivalent amount of petroleum for transportation and residential needs. Both of these objectives have been the subject of active research. With the depletion of the world's petroleum reserves (which will happen) and the increasing dependence on imported petroleum, an efficient coal conversion scheme is a very important long-term goal.

4.1 HISTORICAL OVERVIEW

The preparation of gaseous and liquid fuels from coal is far from new. As early as 1812 London, England, used gas manufactured by the pyrolysis of coal for street lighting. This use spread, and soon most of the larger towns had their own "gasworks" for the production and distribution of "coal gas," or "town gas."

The earliest work on coal liquefaction by direct hydrogenation was started around 1912 by F. Bergius in Germany (1). Bergius' first attempt at large-scale coal liquefaction began in 1916 as a part of the German war effort in World War I. By 1926 Bergius was hydrogenating coal on a 100-kg scale. In 1931 Bergius was awarded the Nobel prize for his work on high-pressure hydrogenation.

During the same time period, F. Fischer and H. Tropsch were experimenting on the synthesis of hydrocarbon fuels from carbon monoxide and hydrogen (2). This mixture of gases was produced by the reaction of steam with coal or coke:

$$C + H_2O \longrightarrow CO + H_2$$

$$nCO + (2n + 1)H_2 \longrightarrow C_nH_{2n+2} + nH_2O$$

During the 1930s, plants utilizing Fischer-Tropsch chemistry were constructed in Germany and other countries to produce hydrocarbons from coal.

From 1931 to 1944, the synthetic fuel industry in Germany grew from a capacity of 2.5 million barrels of synthetic oil per year to a high of 25.5 million barrels per year. During World War II these plants supplied the bulk of Germany's fuel needs.

Following the end of World War II, the U.S. Bureau of Mines obtained a high-pressure liquid ammonia plant at Louisiana, Missouri, and converted it to a coal liquefaction test facility using captured German technology. Both Bergius direct hydrogenation and Fischer-Tropsch processes were investigated. The cost of manufacturing synthetic gasoline at that time was estimated by the U.S. Bureau of Mines to be about twice the cost of petroleum-based gasoline (about $.19 per gallon as opposed to $.11 per gallon). The Bureau of Mines closed the Louisiana, Missouri, operation in 1954.

The Arab oil embargo of 1973 created a renewed interest in coal conversion in the United States and in other coal-producing nations. The coal conversion schemes presented in this chapter were developed during that period of coal research. Unfortunately, the modern coal liquefaction schemes have also largely failed to produce synthetic gasoline at or near competitive prices. This economic failure should not lessen the scientific importance of the work or the importance of coal conversion as a long-term goal. In 1955 the South African Coal, Oil, and Gas Corporation (SASOL) started up a plant for the production of gasoline and other products from coal. The operation of this plant is based on Fischer-Tropsch chemistry. A second SASOL plant went into production in 1982, and a third began operation in 1984. An excellent review of the history of the production of synthetic fuels from coal has recently been published by A. N. Stranges (3).

4.2 PROBLEMS AND OBJECTIVES OF LIQUEFACTION

The first chemical problem in coal conversion is the addition of hydrogen to the coal. The hydrogen content of a typical bituminous coal is roughly 5% by weight, while for gasoline the hydrogen content is about 14% by weight. The addition of hydrogen (hydrogenation) presents several problems. First, the cost of hydrogen or hydrogen generation is relatively high. Second, hydrogenation is a rather difficult reaction to carry out and generally requires high temperatures and pressures. The principal problem in coal liquefaction, then, is to adjust the H:C atomic ratio to a more usable value. The H:C atomic ratios for several coals relative to those of some other fuels and pure compounds are shown in Table 4.1.

All coal conversion processes ultimately increase the H:C atomic ratio of the product relative to that of the parent coal. There are, at least in principle, a number of ways in which this may be accomplished. These methods include the removal of carbon (pyrolysis and solvent extraction), the addition of hy-

TABLE 4.1 H:C Ratios of Some Common Fuels and Pure Compounds

Fuel or Compound	H:C Atomic Ratio
Methane	4.0
Natural gas	3.5
Ethane	3.0
Propane	2.7
Butane	2.5
Gasoline	1.9
Petroleum crude	1.8
Tar sands	1.5
Toluene	1.1
Benzene	1.0
High-volatile B bituminous coal	0.8
High-volatile A bituminous coal	0.8
Lignite	0.7
Medium-volatile bituminous coal	0.7
Anthracite	0.3

drogen (direct or indirect, with or without catalyst), and the complete decomposition and reassembly of the atoms in coal (gasification and Fischer-Tropsch process).

In addition to adjustment of the H:C ratio, the heteroatom content must be significantly reduced. The oxygen-, sulfur-, and nitrogen-containing compounds in the coal liquids act as catalytic poisons in the refining process and have a deleterious effect on fuel performance. The mineral matter in coal must also be removed. The mineral matter creates severe problems in pumping and handling the coal-oil slurry in the liquefaction process. The mineral matter still present in the syncrude must be removed before refining and use. We shall consider first the numerous coal liquefaction techniques and then the coal gasification techniques.

4.3 COAL LIQUEFACTION BY PYROLYSIS

The pyrolysis of coal is, at least in principle, the simplest method of producing liquid fuels with simultaneous improvement of the H:C atomic ratio. This process uses a relatively high temperature to drive off the volatile matter from coal, leaving behind a residue of coke or char. Many of the smaller, more volatile molecules within coal are richer in hydrogen than is the parent coal. The nonvolatile residue is therefore richer in carbon. In effect, this process improves the H:C ratio of the product by the removal of carbon.

The mechanism of coal pyrolysis involves two basic processes. First, the smaller molecules trapped within the capillary structure of coal and those not covalently bonded to the coal matrix are evolved. Second, the weaker bonds of the coal matrix are thermally cleaved to generate free radicals. These radicals may be stabilized (and probably volatilized) by the abstraction of hydrogen or by disproportionation (Fig. 3.1), or they couple with each other to yield nonvolatile char.

Typical coal pyrolysis involves heating a sample to roughly 450–750°C. Pyrolysis at the higher temperatures gives higher yields of gaseous products at the expense of liquid products. Pyrolysis at higher temperatures also tends to produce liquids with a higher aromaticity, while lower temperatures tend to favor the production of a product with somewhat more aliphatic and hydroaromatic content. Hydrogenation is generally required to upgrade the coal liquids to usable fuels. The hydrogenation step increases the cost of the pyrolysis process. The major disadvantage of the pyrolytic conversion process is that the yield of liquid product is low, while the yield of by-product char is relatively high. Nevertheless, a number of pyrolysis processes have been developed for potential commercial exploitation.

One of the earlier pyrolytic liquefaction processes to be developed on a large scale is FMC's Char Oil Energy Development project (COED). The COED process is a fluidized-bed process in which coal pyrolysis is carried out at progressively higher temperatures. A fluidized-bed process is one in which the gas flow through the coal particles is sufficient to float them in the gas stream and give the appearance of a bubbling fluid bed. The advantage of using progressively higher temperatures is that the partial devolatilization of the coal in the lower temperature stage prevents the coal from caking or agglomerating at the higher temperatures, which would plug the reactor bed. The COED process is summarized in Figure 4.1 (4,5).

Crushed coal (nominal-⅛-in mesh size) is dried and carried by the fluidizing gas into the pretreater (stage 1). Stage 1 is maintained at around 177°C for the lower-rank coals and up to about 343°C for the bituminous coals. This is below the temperature at which active thermal decomposition of coal takes place and minimizes the caking, or agglomerating, character of the coal. The coal then flows into stage 2, where it is pyrolyzed at 427°C by hot recycle flue gas. The coal then flows into stage 3, where additional pyrolysis occurs by means of hot recycle gas and hot char (recycled from stage 4). This stage is typically maintained around 538°C. The coal then moves to stage 4, where the final pyrolysis takes place at 816°C.

The raw COED oil product is removed from the reactors as a vapor, leaving the residual char to be withdrawn as a fluidized solid. Part of the by-product char is burned to provide heat for the process. Any solid material contained in the condensed oil is removed by a filtration process. The raw COED oil has a relatively low H:C ratio, a high concentration of oxygen, a

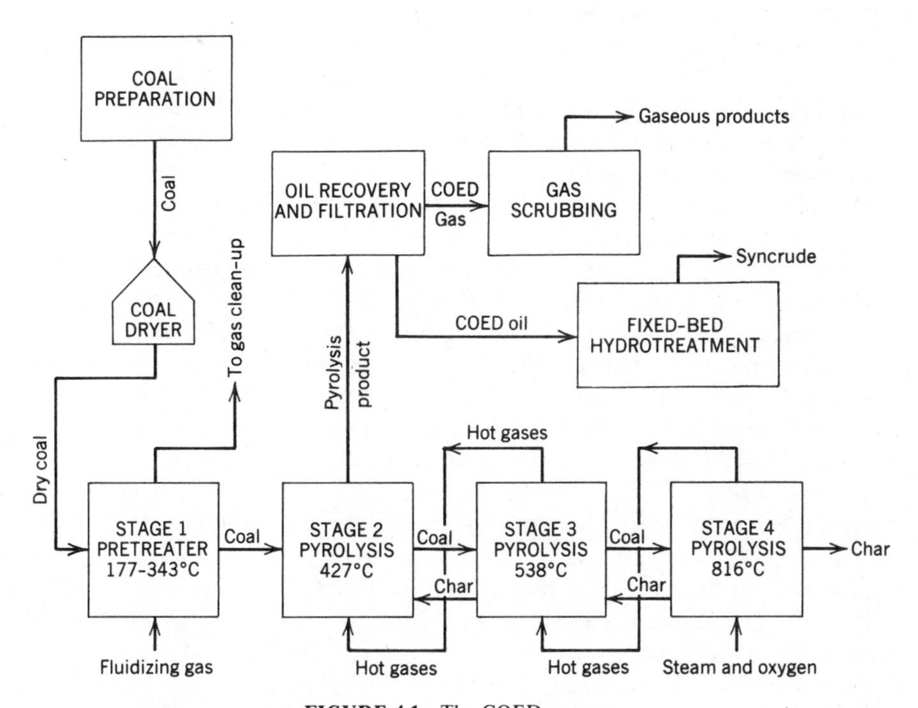

FIGURE 4.1 The COED process.

high sulfur content, and a relatively high nitrogen content (6). The solid-free, raw oil must, therefore, be upgraded by hydrogenation. The raw COED oil is typically hydrogenated at 340–450°C under a hydrogen pressure of around 21 MPa. The reaction is carried out in a fixed-bed reactor using a catalyst consisting of 3% NiO and 15% MoO_3 on alumina. This treatment reportedly removes approximately 72% of the nitrogen, 91% of the oxygen, and about 98% of the sulfur (6). This process yields approximately 1.5 barrels of oil per ton of bituminous coal. Yields are generally lower for lower-rank coals. The COED process has been modified to gasify the by-product char. The combined process to produce liquid fuels and gaseous fuels is called the COGAS process (7).

The TOSCOAL process has been developed by the Tosco Corporation for the low-temperature pyrolysis of coals of low rank (lignite and subbituminous) (8,9). Coals of higher rank must be pretreated (mildly oxidized) to prevent caking, or agglomeration, before they can be used as feed for the TOSCOAL process. The preheated coal is fed into a horizontal rotating drum reactor, where it is heated to 427–534°C in contact with ceramic balls. Process heat is provided by the hot ceramic balls, which are heated in a separate furnace fired by product gas or char. The pyrolysis vapors are collected and condensed to yield the raw TOSCOAL oil. The gas produced by the process may be used as fuel for the ball furnace. The by-product char has potential for use as boiler fuel or as feed for gasifiers. As with the COED process, the raw TOSCOAL oil must be upgraded before use as a fuel. The overall product yield is 0.3–0.5 barrels of oil per ton of coal.

When coal is heated to the active decomposition temperature very rapidly (flash pyrolysis), it tends to undergo more extensive fragmentation (10). This leads to enhanced product yields. Several schemes have been tested for the flash pyrolysis of coal. One of the most advanced is the Lurgi-Ruhrgas flash carbonization process. This technology was developed by Lurgi GmbH, a subsidiary of Metallgesellschaft AG, Frankfurt, and Ruhrgas AG, Germany (11). High heating rates are realized by intimately mixing the feed coal and recycled hot char in a mechanical mixer. The product vapors are withdrawn at the end of the mixer, passed through a hot cyclone to remove dust, and fed to a condenser unit. The residence time of the product in the pyrolysis zone is very short (in the order of seconds). This minimizes secondary thermal decomposition of the product and results in higher overall yields. An optimum temperature of 450–600°C is employed, the exact temperature depending on the rank of coal used. The intimate mixing of the coal and hot char overcomes the agglomeration problems and permits the use of both caking and noncaking coals in the Lurgi-Ruhrgas process.

The Garrett flash pyrolysis process (also called the Oxy flash pyrolysis process or the Occidental flash pyrolysis process) also employs hot recycled char to provide process heat (12,13). The process is flexible in that it can be set up to produce either liquid products (oil) or a hydrocarbon-rich gas. The pulverized feed coal is mixed with hot char and fed into an entrained-flow reactor.

An entrained-flow reactor is one in which a carrier gas of sufficient flow rate is used to suspend or entrain the coal particles in the gas stream. A temperature of approximately 580°C is employed for oil production and one of about 927°C for gas production. The product is passed through cyclones to remove dust and then subjected to further processing. A portion of the by-product char is recycled, during which some of it is burned with air to provide process heat. The hot char is then recycled to the pyrolysis reactor.

Hydrogen has been employed as the carrier gas in entrained-flow reactors in an attempt to improve product yield. Such flash hydropyrolysis experiments have been carried out on a small scale. The Rockwell flash hydropyrolysis process serves to illustrate the technique (14). This process has been developed by Rockwell International under sponsorship of the U.S. Department of Energy. The Rockwell process is designed to very rapidly mix, react, and quench the coal. The mixture of pulverized coal and hot hydrogen is fed into an entrained-flow reactor at approximately 980°C and 3.5–10 MPa pressure. The coal and hydrogen react for approximately 0.1 s, after which time the reactor's effluent is quenched by water spray to minimize further hydrogenation of the initial liquid products. The product yields from hydropyrolysis experiments are superior to normal pyrolysis yields, but further refinements must be made to justify the additional cost.

4.4 DIRECT HYDROLIQUEFACTION OF COAL

The direct hydroliquefaction of coal has been a very widely studied process. Most of the current processes for the liquefaction of coal have features in common and are largely extensions of the early work by Bergius (1). In the majority of the conversion processes, the coal is dried and pulverized, and then slurried with a coal-derived recycle oil. The slurry is then pumped into a high-pressure reactor, where the coal is liquefied by hydrogenation at high temperature and pressure. The operating temperature and pressure are similar for most of the liquefaction processes, as can be seen in Table 4.2.

Under mild conditions (low temperature, low pressure, and short reaction times) or in the absence of a catalyst, the liquefaction product is usually a rather heavy oil suitable for boiler fuel applications. Under more severe conditions, the heavy oil is further converted into distillable products. In either case, the products generally require further upgrading before they are suitable for use as fuels.

The operating temperature (approximately 450°C) provides an important clue to the mechanism most likely involved in the coal liquefaction process. It has been reasonably well established that above 400°C the pyrolytic breakup of the coal matrix begins. It is very likely that a large number of concurrent and competitive chemical reactions occur during the liquefaction process. Thermolysis, hydrogen abstraction, dealkylation, cleavage of bridges between structural units, desulfurization, dehydration, and ring opening are but a few

TABLE 4.2 Some Representative Liquefaction Conditions

Process	Temperature (°C)	Pressure (MPa hydrogen)	Catalyst
Bergius	480	30–70	Iron oxide
U.S. Bureau of Mines SYNTHOIL	450	14–28	Cobalt moly[a]
Ashland H-coal (H-coal)	450	19	Cobalt moly[a]
Exxon donor solvent (EDS)	450	10	None[b]
Gulf solvent refined coal (SRC-II)	460	13	None[b]
Dow coal liquefaction	460	14	Emulsion[c]
CO-STEAM	450	20–30	None[b]
Conoco zinc chloride	415	21	Zinc chloride

[a] Employs a commercial pelletized cobalt molybdate on alumina.
[b] While no catalyst is added to the liquefaction reactor, the mineral matter in coal is reported to exhibit mild catalytic activity.
[c] A proprietary emulsion catalyst is employed.

of the many possibilities. Four of the possible reactions are illustrated in Figure 4.2. Some of the more likely candidates for bond thermolysis were discussed in Section 3.1.

Once one or more of these weaker bonds has been cleaved, the free radicals generated can be stabilized by the addition of hydrogen. The hydrogen may be added either from molecular hydrogen or from a hydrogen-donor solvent, especially when a good donor solvent (such as tetralin or a tetralinlike species) is present:

$$\xrightarrow{-H} \qquad \xrightarrow{-3H}$$

Tetralin or tetralinlike molecules are good donors due to the fact that the free radical intermediate is a relatively stable benzylic-type free radical and the product after the loss of four hydrogen atoms is a stable aromatic compound (naphthalene). In the absence of a good hydrogen source, the free radicals may couple to form coke or char. The importance of hydrogen is schematically illustrated in Figure 4.3.

The hydrogen transfer reactions appear to be purely thermal processes (15). Attempts to accelerate the reaction by the use of contact catalysts have been largely unsuccessful. Coal itself is, of course, not in solution during the initial phase of the liquefaction process. This fact would seem to preclude the possibility of catalysis in the initial phase of coal liquefaction. It is much more likely that the donor solvent (or molecular hydrogen) rapidly diffuses into or dissolves in the coal particle and that the hydrogen transfer reaction takes

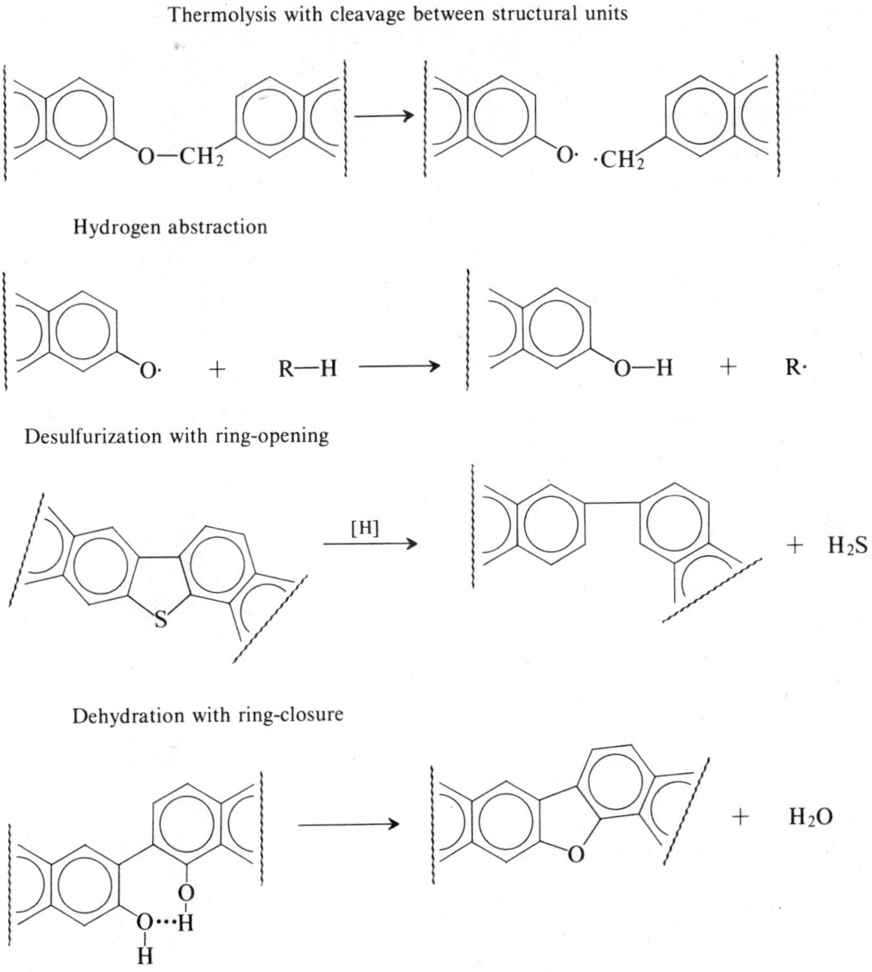

Thermolysis with cleavage between structural units

O—CH_2 → O· ·CH_2

Hydrogen abstraction

O· + R—H → O—H + R·

Desulfurization with ring-opening

S [H] → + H_2S

Dehydration with ring-closure

O O···H H → O + H_2O

FIGURE 4.2 Some possible reactions during hydroliquefaction.

place between the solvent (or hydrogen) and the radical species within the coal particle itself. The function of a donor solvent, then, is to enter the coal particle and provide a source of hydrogen for the free radicals generated thermally within the matrix. The rate of hydrogen transfer with the more active donors appears to be roughly the same regardless of the structure of the donor (15). This indicates that it is the rate of thermal decomposition of the coal matrix that actually determines the extent of hydrogen transfer once a sufficiently reactive donor is available.

S. R. Gun and coworkers have studied the hydrogenation of coal using both molecular hydrogen and donor solvents (16,17). The results, shown in Table 4.3, indicate that hydrogenation by molecular hydrogen and by donor

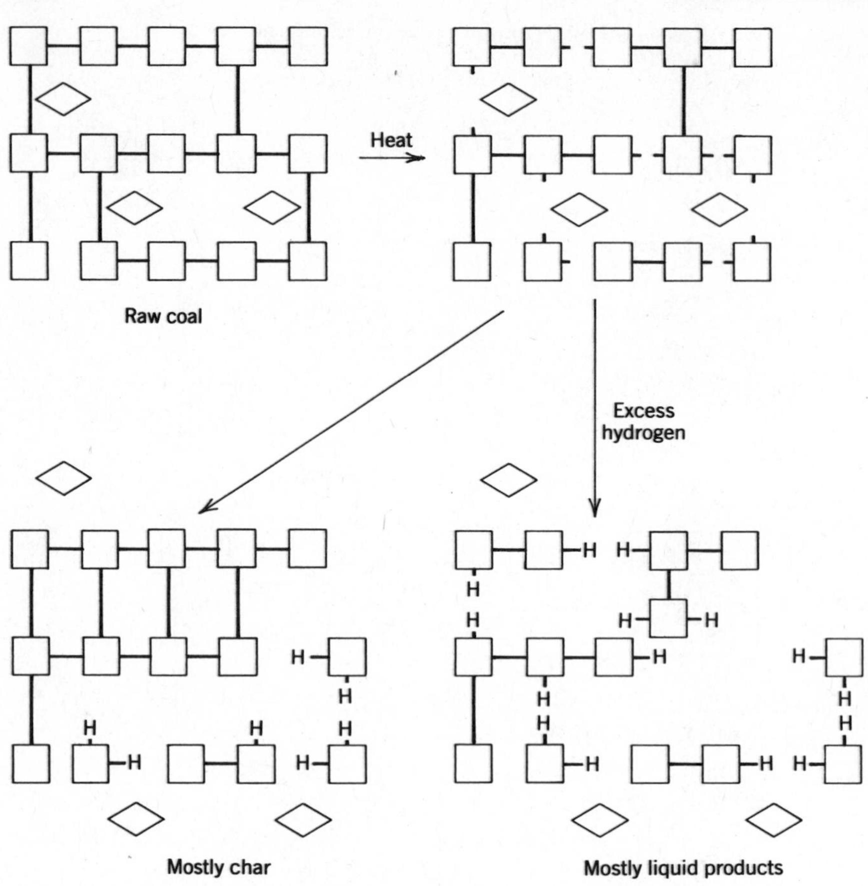

FIGURE 4.3 Importance of hydrogen in coal liquefaction.

solvent are both important. In fact, it appears that both are roughly equal in their ability to hydrogenate the free radicals formed in the thermolysis of the coal matrix. Molecular hydrogen in the presence of a catalyst yielded a 75.3% conversion of the coal into volatile (gaseous plus benzene-soluble) products. Under identical conditions, the donor solvent tetralin, even in a nitrogen atmosphere, increased the conversion yield from 16.3% to 66.3%. The highest conversion is obtained using a combination of molecular hydrogen, a catalyst, and a good hydrogen donor. These results support the view that the conversion process is free radical in nature, with liquefaction ultimately achieved by stabilization of the coal fragments by the addition of hydrogen from either molecular hydrogen or donor solvent.

The pyrolysis of 1,2-diphenylethane at 450°C provides some interesting information about the differences between molecular hydrogen and a donor solvent in the stabilization of thermally generated free radicals (18). What

TABLE 4.3 Hydrogen versus Tetralin as Donor

Gas Pressure (MPa)	Added Tetralin (g)	Added Catalyst (g)	Conversion (wt% coal)
9.8 (nitrogen)	—	—	16.3
9.8 (nitrogen)	—	0.25	16.9
9.8 (nitrogen)	50.0	—	66.3
9.8 (nitrogen)	50.0	0.25	79.4
9.8 (hydrogen)	—	—	46.6
9.8 (hydrogen)	—	0.25	75.3
9.8 (hydrogen)	50.0	—	75.4
9.8 (hydrogen)	50.0	0.25	96.7

Source: References 16 and 17; reprinted with permission from S.R. Gun et al., "A Mechanistic Study of Hydrogenation of Coal, Part 2," *Fuel, 58,* 176, 1979, Butterworth & Co., Publishers, Ltd.)

Note: The coal used was 79.9% C, 5.9% H, 4.4% S, 1.3% N, 8.5% O, and 43.8% volatile matter; 25 g of coal were heated to 400°C for 1 h. The catalyst was ammonium molybdate impregnated in the coal.

products are obtained depends on whether a source of hydrogen is available and the form of the hydrogen source (molecular hydrogen or tetralin). Pyrolysis in the absence of a hydrogen source yields toluene and stilbene as the major products. In the presence of a good hydrogen donor (such as tetralin), the only product isolated is toluene. Pyrolysis in the presence of molecular hydrogen yields a mixture of toluene, benzene, and ethylbenzene:

$$C_6H_5{-}CH_2CH_2{-}C_6H_5 \xrightarrow{450°C} C_6H_5CH_3 + C_6H_5{-}CH{=}CH{-}C_6H_5$$

$$C_6H_5{-}CH_2CH_2{-}C_6H_5 \xrightarrow{450°C\ [\text{tetralin}]} 2\ C_6H_5CH_3$$

$$C_6H_5{-}CH_2CH_2{-}C_6H_5 \xrightarrow{450°C\ [H_2]} C_6H_5CH_2CH_3 + C_6H_5CH_3 + C_6H_6$$

The observations from the pyrolysis of 1,2-diphenylethane can be explained in terms of the mechanism shown in Figure 4.4. The initial reaction involves thermolytic cleavage to form the benzyl radical (reaction 1). In the absence of

1. $C_6H_5—CH_2CH_2—C_6H_5 \xrightarrow{450^\circ C} 2C_6H_5—CH_2^{\cdot}$

2. $2C_6H_5—CH_2^{\cdot}$ + Tetralin $\longrightarrow 2C_6H_5—CH_3$ + Dihydronaphthalene

3. $C_6H_5—CH_2^{\cdot} + C_6H_5—CH_2CH_2—C_6H_5 \longrightarrow C_6H_5—CH_3 + C_6H_5—CH{=}CH—C_6H_5$

4. $C_6H_5—CH_2^{\cdot} + H_2 \longrightarrow C_6H_5—CH_3 + H^{\cdot}$

5. $C_6H_5—CH_2CH_2—C_6H_5 + H^{\cdot} \longrightarrow C_6H_5—CH_2CH_3 + C_6H_6$

FIGURE 4.4 Mechanism for the pyrolysis of 1,2-diphenylethane. (*Source:* Reprinted with permission from L.W. Vernon, "Free Radical Chemistry of Coal Liquefaction," *Fuel, 59,* 102, 1980, Butterworth & Co., Publishers, Ltd.)

any added hydrogen source, the benzyl radical is stabilized by hydrogen abstraction from 1,2-diphenylethane (reaction 3), thus yielding toluene and stilbene. In the presence of added tetralin, the benzyl radical is quickly converted to toluene (reaction 2). In the presence of molecular hydrogen, the abstraction of a hydrogen by the benzyl radical results in the formation of atomic hydrogen (reaction 4). The hydrogen atom, being very reactive, is believed to promote cleavage of 1,2-diphenylethane to form benzene and ethylbenzene (reaction 5). These experiments indicate that different product distributions can result depending on whether molecular hydrogen or a donor solvent is used to quench the thermally generated free radicals. The products from the hydroliquefaction of coal may, therefore, depend to some extent on the kind of hydrogen source employed.

4.5 COAL-DERIVED LIQUIDS

The liquid products of coal conversion are classified according to their solubility behavior. The liquids are classified as oils (soluble in pentane or hexane), asphaltenes (soluble in benzene or toluene, insoluble in hexane), and preasphaltenes (soluble in pyridine, insoluble in benzene or hexane). That material which is insoluble in pyridine is usually classified as residue, or mineral matter.

The course of the coal liquefaction process is believed to occur in a step-wise manner (19,20). Data indicate that initially there is a rather rapid thermal cleavage of a few of the most reactive bonds to produce relatively large organic fragments that may be classified as preasphaltenes. These compounds contain a fairly large amount of oxygen in the form of phenolic groups (21). In the second step of the process, the somewhat more resistant bridges are cleaved to form the yet smaller fragments, which may be classified as asphaltenes. Finally, there is a rather slow reaction in which the asphaltenes appear to be converted into materials, which are classified as oils.

It should be noted, however, that some oils are formed very rapidly in the initial stages of the reaction. These materials may well be those largely hydrocarbon substances trapped within the pore structure (or noncovalently bonded to the coal matrix), which are released upon decomposition of the coal matrix. Activation energies support this view. Studies of activation energies seem to indicate that two different processes occur during the collapse of the coal matrix (21–24). The combination of a high rate of conversion and a relatively low activation energy (a maximum of approximately 30 kcal/mole) suggests that the initial rapid process likely involves either physical processes or the cleavage or destruction of nonvalence bonds (i.e., hydrogen bonds). The activation energy for the somewhat slower second process is still rather low, at approximately 45–60 kcal/mole. This activation energy suggests the thermolytic cleavage of rather weak covalent bonds with the formation of stable free radicals. These data are consistent with the view that a part (10–25% by weight) of the material from coal conversion results from noncovalently bonded species, while the bulk of the material results from the fragmentation of the coal matrix.

The conversion of asphaltenes to oils is a much slower process, and obviously much stronger bonds are involved. High yields of oils generally require more severe reaction conditions (higher temperature and pressure, and longer residence times). Some of the reactions that are possibly involved in the conversion of asphaltene to oil include hydrogenation of aromatic rings, dehydration, loss of heteroatoms by ring opening of heterocyclic rings, and cleavage of bridging structures. The stepwise nature of the coal conversion process may be represented as in Figure 4.5. The average molecular weights are included for comparison purposes only. Number-average molecular weight measurements of coal-derived liquids tend to be unreliable and are generally somewhat low.

Every coal liquefaction process yields an extremely complex mixture containing hundreds of compounds. Some of these are desirable components in a fuel or feedstock, and some are very undesirable. In general, the desirable compounds are those classified as oils (hexane soluble). Hexane will, for example, dissolve most hydrocarbons but relatively few phenolics and polar

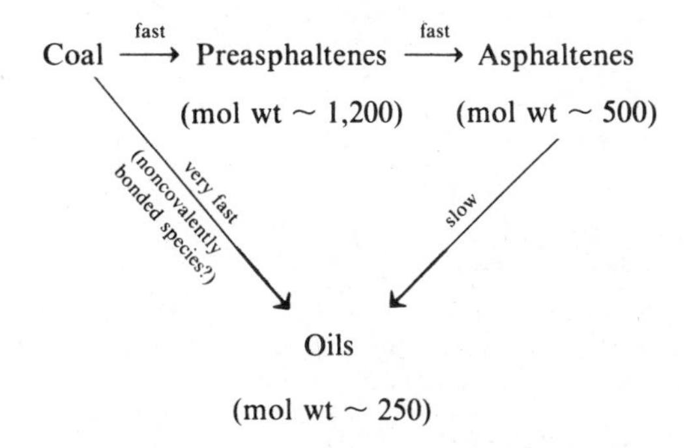

FIGURE 4.5 Stepwise nature of the conversion process.

heterocyclics. Most of the undesirable compounds fall into the asphaltene and preasphaltene categories. Most of the coal liquefaction products fall into the oil and asphaltene classes. The oil fraction has a lower number-average molecular weight (250–400) than do the asphaltenes (500–700) and generally consists of two kinds of compounds (25–27). There is the larger hydrocarbon fraction (approximately 80%), consisting of both aliphatics and aromatics, and a smaller fraction (approximately 20%) rich in phenols and heterocyclics.

The coal-derived asphaltenes appear to have some acid-base structure and may be separated by chromatographic procedures into neutral, acidic, and basic fractions (28). The acidic component is largely characterized by derivatives of phenols. The basic component consists mostly of heterocyclic and ether oxygen and pyridine derivatives. As a rule, the asphaltene fraction is predominantly aromatic with a relatively large heteroatom content.

Because of the high degree of acid-base functionality and polarity, the asphaltenes exert a very large effect on the viscosity of the coal-derived liquids. The viscosity appears to increase exponentially with asphaltene content (29). The molecular weight of the asphaltenes is also directly related to the viscosity of the coal liquid. This resultant high viscosity necessitates some sort of additional processing before transportation and storage of the liquid product.

The high heteroatom content also makes further processing a necessity before ultimate conversion into fuels. If not removed, the nitrogen and sulfur would result in high levels of emission of their oxides when these materials were burned as fuels. The sulfur content is much more easily removed than is the nitrogen content. Even those coal conversion processes that use no added catalyst remove most of the organic sulfur as well as most of the oxygen and inorganic sulfur. The bulk of the nitrogen, however, remains in the coal-derived liquids. The coal conversion reactions that use a hydrogenation catalyst appear to be even more effective in the removal of sulfur from the coal-derived liquids. This is partially due to the fact that the hydrogenation catalysts are also effective hydrodesulfurization catalysts:

$$\text{(dibenzothiophene)} \xrightarrow[\text{catalyst}]{[H]} \text{(biphenyl)} + H_2S$$

Nitrogen, however, is still a problem. Hydrogenation catalysts are not effective hydrodenitrogenation catalysts. The liquids derived from coal are potentially valuable as petroleum substitutes and feedstocks, but refining operations, such as cracking and hydrocracking, require a feedstock that is very low in nitrogen. The basic nitrogen compounds found in coal liquids are very effective catalytic poisons and thus must be removed.

It is interesting to compare the properties of a typical coal-derived syncrude with those of a representative normal petroleum-based crude. The first notable difference is aromaticity. The coal-derived liquids are about 70% aromatic. This aromaticity is much higher than that of petroleum crude. The sulfur content of coal-derived syncrude is roughly 0.3–3%, the oxygen content is 1–10%, and the nitrogen content is about 1–1.3% by weight. Petroleum crude, on the other hand, contains approximately 0.2% nitrogen, about 1.2% sulfur, and normally little or no oxygen. Since one of the objectives of all liquefaction procedures is to produce a clean-burning, low-sulfur fuel, it appears that the first step in upgrading coal-derived liquids must be a severe catalytic hydrogenation to remove the bulk of the heteroatoms and to convert the heavy, viscous oil to a lighter distillable liquid (basically, to convert the asphaltenes to oils).

It should be recognized that raw coal-derived liquids contain many toxic and potentially hazardous compounds (30). Coal liquefaction products and by-products contain compounds that have been shown to be carcinogenic (31). Some of the more common examples are polynuclear aromatic hydrocarbons and their alkyl derivatives. Exposure to coal-derived liquids and their vapors should always be considered hazardous and special precautions always observed.

4.6 CATALYSIS IN COAL LIQUEFACTION

Before discussing catalysis in coal liquefaction, one should understand the role of the catalyst in a chemical reaction. A catalyst is a substance that alters the rate of a chemical reaction without being used up in the process. Most catalysts are used to accelerate the rate of reaction. The energy of activation for a chemical process is an energy barrier, or "hill," that must be overcome if the reactants are to be successfully converted into products. One effect of a catalyst is to lower the energy of activation, or the hill, that must be overcome. In general, the lower the energy of activation, the greater the number of molecules that can overcome the barrier and react in a given time. A catalyst need only have a mild effect on the reaction to be commercially important. A small decrease in operating temperature, a small decrease in operating pressure, or a small increase in yield or efficiency can have a large impact on the overall economics of a large-scale process.

Catalysts are frequently divided into homogeneous and heterogeneous types. In homogeneous catalysis the catalyst is present in the same phase as the reactants, while in heterogeneous (or surface) catalysis the catalyst and reactants exist in different phases. In heterogeneous catalysis the reactions occur on the catalyst surface.

Even though the term *catalytic liquefaction of coal* has been used, it is clear that the initial breakup of the organic matrix in coal cannot be accelerated by adding a solid, since catalysis requires intimate contact, which cannot be achieved until some initial liquid product has formed. There is evidence that first there is a thermolytic breakup of the matrix (which is accelerated by adding neither a catalyst nor a hydrogen source) followed by hydrogen transfer to the resultant free radicals (32). However, the catalyst is very helpful in upgrading the donor species present and in such processes as hydrodesulfurization, hydrodeoxygenation, and reduction of some of the more reactive unsaturated compounds present. Exxon, for example, performs a separate catalytic hydrogenation of their recycle oil to improve its hydrogen donor characteristics ("solvent quality index"). The reactions between coal, donor solvent, hydrogen, and catalyst are illustrated in Figure 4.6.

A large number of catalysts have been tested for coal liquefaction and coal liquid upgrading, yielding varied results (33). Without question, the least expensive catalyst is the mineral matter in the coal itself. The catalytic activity of the mineral matter has been investigated and used in the Gulf solvent refined coal process and the Exxon donor solvent process (34–36).

Transition metal complexes of cobalt, ruthenium, rhodium, palladium, and platinum have been tested as homogeneous catalysts in the hydrogenation of coal and coal-derived liquids (37). Complexes such as dicobalt octacarbonyl [$Co_2(CO)_8$] have been shown to be effective for the hydrogenation of polynuclear aromatics and coal (38). N. Holy and coworkers tested the rhodium complex of N-phenylanthranilic acid for coal liquefaction (39). The major problem with all catalysts, especially homogeneous catalysts, is their recovery from the product mixture. Since these catalysts are extremely expensive, anything less than virtually total recovery would render their use prohibitively expensive. Metals such as platinum are very effective hydrogenation catalysts, but the presence of heteroatoms, especially sulfur, in coal and coal liquids very quickly poison the catalyst and render it ineffective.

Zinc chloride ($ZnCl_2$) has been very successfully used as a hydrocracking catalyst for coal and coal liquids (40,41). Hydrocracking is a technique by which smaller organic molecules are produced by the thermal cleavage (pyrolysis) of larger species in the presence of hydrogen and a catalyst. The use of zinc chloride results in a more rapid conversion at lower temperature (around 400–425° C) than does the use of any conventional catalyst tested to date (42). The use of zinc chloride also results in a more complete conversion of the coal and reduction of the heteroatom content, and yields products with potentially higher octane rating than do other, more conventional catalysts (43).

The problems associated with the use of zinc chloride are related to the massive amounts of catalyst that are generally required (1:1 catalyst to coal ratio) and to the fact that this catalyst is difficult to recover from the product stream in sufficient yield to be very cost effective. During the hydrocracking of coal, some of the zinc chloride is converted to zinc sulfide (from the sulfur present) and to an ammonia-zinc chloride complex (from the nitrogen present). These compounds must be converted back to zinc chloride and returned to the reactor (44). Significant engineering problems are also associated with handling the massive amounts of molten, corrosive zinc chloride.

The catalyst most commonly used for coal liquefaction at the present is a sulfided cobalt molybdate supported on alumina. This catalyst is active for hydrogenation of the materials evolved in the depolymerization of the coal matrix and is reasonably long-lived in the presence of heteroatoms in the coal-derived liquids. There is still a serious problem with the mineral matter, which clogs the pores of the catalyst surface and renders it inactive. The catalyst is frequently used in conjunction with a recycle oil (coal-derived), which is also a hydrogen donor. The use of a combination of hydrogen, donor solvent, and catalyst appears to give the best overall yield of liquefaction product (17). The donor solvent gives up hydrogen to the free radicals from the fragmenting coal matrix and then returns to the catalyst surface to react with the dissociatively adsorbed hydrogen and become "renewed." The renewed solvent may then return to the coal matrix to again donate hydrogen to the radical fragments. Thus the recycle, or donor, solvent functions as a hydrogen carrier, and the catalyst is used to renew or upgrade the solvent. This is illustrated in Figure 4.6. In its donor solvent process, Exxon upgrades the recycle solvent in a separate step outside the liquefaction reactor.

4.7 COAL REACTIVITY IN LIQUEFACTION

It has been well documented that the overall reactivity of coal toward liquefaction is dependent on the recycle solvent, the reaction conditions, and the composition of the coal itself. It would be very desirable to find some parameter or set of parameters that would correlate well with the reactivity or degree of conversion under a given set of reaction conditions. It is important to understand how specific coal characteristics affect coal liquefaction behavior, since such knowledge would be useful in the selection of candidate coals for liquefaction and could aid in the selection of sites for the construction of liquefaction facilities (i.e., near a large reserve of coal with high liquefaction reactivity). The heterogeneity and variability of coal as a feedstock makes difficult any truly quantitative assessment of reactivity. A large number of parameters have been tested with varying degrees of success. These include the percent carbon (rank), the H:C atomic ratio, the volatile matter content, the percent of reactive macerals in the coal, the mineral matter in coal, and the elemental composition.

The percent carbon in coal has been used with limited success as a parameter in predicting the extent of coal conversion. The data do indicate lower yields of oil from the lignite and subbituminous coals than from coals of higher rank. However, the conversion yield falls off rather quickly, at the upper end of the bituminous range and is very low for coals with greater than 90% carbon (dry, ash-free basis) (45). Coals of the high-volatile bituminous rank (82–84% dry, ash-free carbon) appear to give the best liquefaction yields. Within the bituminous range, however, there is poor correlation between rank and conversion yield.

The relationship between the conversion yield and the H:C atomic ratio for a sample of coal does correlate quite well under certain limited conditions of

Coal

pyrolysis

+

+

H_3C

OH

Liquefaction/hydrogenation products

hydrogen

transfer

Hydrogen
carrier
cycle

catalytic
hydrogenation

catalytic
hydrogenation

H_2

Dehydrogenated catalyst
surface

Hydrogenated catalyst
surface

Hydrogenated catalyst
surface

Dehydrogenated catalyst
surface

H_2

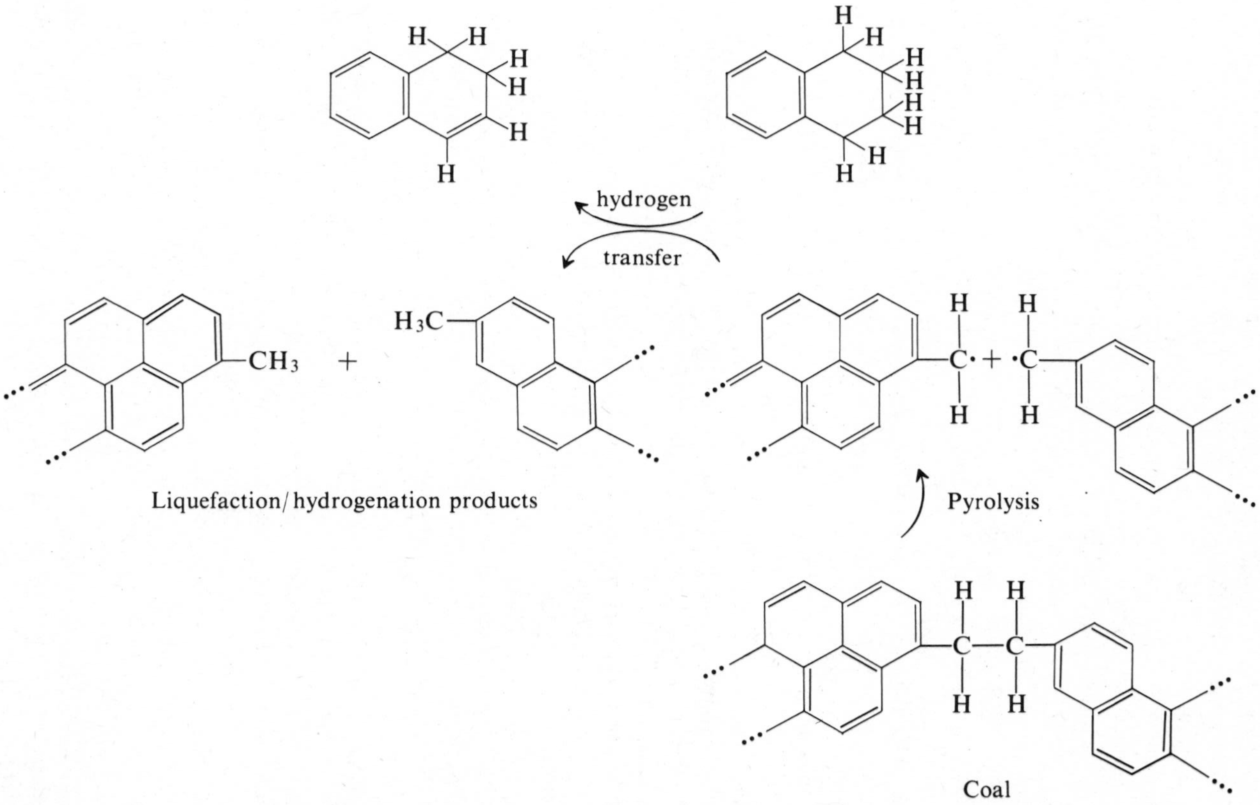

hydrogen transfer

Liquefaction/hydrogenation products

Pyrolysis

Coal

FIGURE 4.6 Reactions between coal, donor solvent, hydrogen, and catalyst. (*Source:* Reprinted with permission from B.C. Gates, "Liquefied Coal by Hydrogenation," *CHEMTECH*, *9*(2), 99 (1979).

liquefaction. Studies of Australian coals, for example, showed that the conversion yield systematically increased as the H:C atomic ratio increased over the range from 0.60 to 0.85 (46). The relationship has not, however, been shown to be general for the bulk of American coals.

The relationship between the percent volatile matter in a sample of coal and the conversion yield has been found to be highly dependent on the actual reaction conditions as well as on the nature of the coal itself (47). However, it is safe to say that coals of higher volatile matter are more reactive than are those of lower volatile matter. Medium-volatile bituminous and higher-rank coals are known to have rather poor conversion yields.

A good relationship between the petrographic composition and conversion yield has been found in many cases (48). The parameter used for petrographic composition has ranged, however, from maximum vitrinite reflectance to total reactive macerals. It is well known that macerals such as vitrinite and exinite are converted into liquid products in high yields on hydrogenation (49). Fusinite remains largely unconverted on attempted liquefaction. The vitrinite and exinite macerals are therefore commonly called the reactive macerals, while fusinite is generally referred to as an inert maceral. For coals with carbon contents between 75% and 82%, there is a very good correlation between the percent of reactive macerals in the coal and the liquefaction yield. As before, the conversion yield is found to decrease very rapidly between 88% and 90% carbon content.

The mineral matter content of a sample of coal often has quite a large effect on the conversion yield during liquefaction. However, the precise effect of mineral matter on coal conversion is not well understood. When pyrite is present in coal, it is converted to pyrrhotite under liquefaction conditions and can act as a hydrogenation catalyst (50). Exactly how well the mineral matter functions depends on how it is dispersed in the coal. Excessive mineral matter contributes, of course, to such problems as catalyst deactivation and reactor plugging.

Plotting the conversion yield against the total sulfur content for a large number of coals shows a rather good correlation (51). The conversion yield has a strong tendency to increase with increasing sulfur content for most of the coals studied. The usefulness of any given parameter in predicting conversion yield can be expressed in terms of a correlation coefficient. The correlation coefficient r is a measure of linear association or dependence. If all points fall on a straight line not parallel to one of the coordinate axes, r has a value of 1 or -1. This indicates a perfect positive or perfect negative correlation. For example, one study showed that the percent carbon in coal had a correlation coefficient (with conversion yield) of $-.52$, the total reactive macerals had a correlation coefficient of .80, and the total sulfur content had a correlation coefficient of .93.

Given and coworkers have reported that the highest correlation coefficients are obtained by dividing the coal into three separate groups for analysis (52). Group 1 contains high-rank, medium-sulfur coals; group 2 contains medium-to-low-rank, high-sulfur coals; and group 3 contains medium-to-low-rank, low-sulfur coals. The best correlation for group 1 coals is obtained using a combination of vitrinite reflectance, H:C ratio, and total vitrinite content. The best parameters for group 2 include volatile matter, vitrinite reflectance, and total sulfur content. Group 3 shows the best correlation with volatile matter and total reactive maceral content. It should be possible, therefore, with proper choice of coal parameters, to estimate a given coal's liquefaction behavior under a predefined set of operating conditions.

4.8 COAL LIQUEFACTION PROCESSES

4.8.1 The SYNTHOIL Process

A number of coal liquefaction schemes have been developed for possible commercial exploitation. The U.S. Bureau of Mines (now a part of the U.S. Department of Energy) SYNTHOIL process was one of the earlier direct hydroliquefaction processes to be developed in the United States, with initial work starting in 1969 (53,54). This is one of four processes discussed here in which coal hydrogenation is carried out in the presence of an added catalyst. A simplified flowsheet for the SYNTHOIL process is shown in figure 4.7.

The coal is crushed (70% through 200 mesh) and dried. The coal is then slurried with a coal-derived recycle oil (35–40% coal), mixed with hydrogen, and fed into a preheater. The slurry is pumped with hydrogen into a fixed-bed catalytic reactor containing pelletized cobalt molybdate on alumina. The slurry is forced through the catalyst bed with sufficient velocity to keep the bed agitated. This turbulence provides more effective contact between the three phases (catalyst-coal, recycle oil, and hydrogen) and aids in keeping the mineral matter from deactivating the catalyst. As mentioned earlier, catalyst deactivation occurs when mineral matter collects in the catalyst pores. The effluent from the reactor is cooled, and the liquids and unreacted solids are separated from the gases. The gases are fed into a gas purification system, where hydrogen is removed and recycled. The hydrocarbons are also separated from water, ammonia, and hydrogen sulfide. The liquids and solids are then separated in a centrifuge. From the centrifuge, part of the oil is removed as product, and part is recycled as slurry vehicle. The solids may then be pyrolyzed to yield additional liquid product and the residue gasified to produce hydrogen.

The SYNTHOIL reactor typically operates around 450° C with a hydrogen pressure of 14–28 MPa. Tests have also been carried out without the cobalt molybdate catalyst. The mineral matter in coal is known to have some catalytic ability, and coal is successfully liquefied without added catalyst. The SYNTHOIL process has operated at the 0.5 ton per day level, and typical oil yields range from 3.0 to 3.3 barrels per ton of bituminous coal. The crude coal-derived oil must be upgraded by hydrogenation before use as a fuel or feedstock.

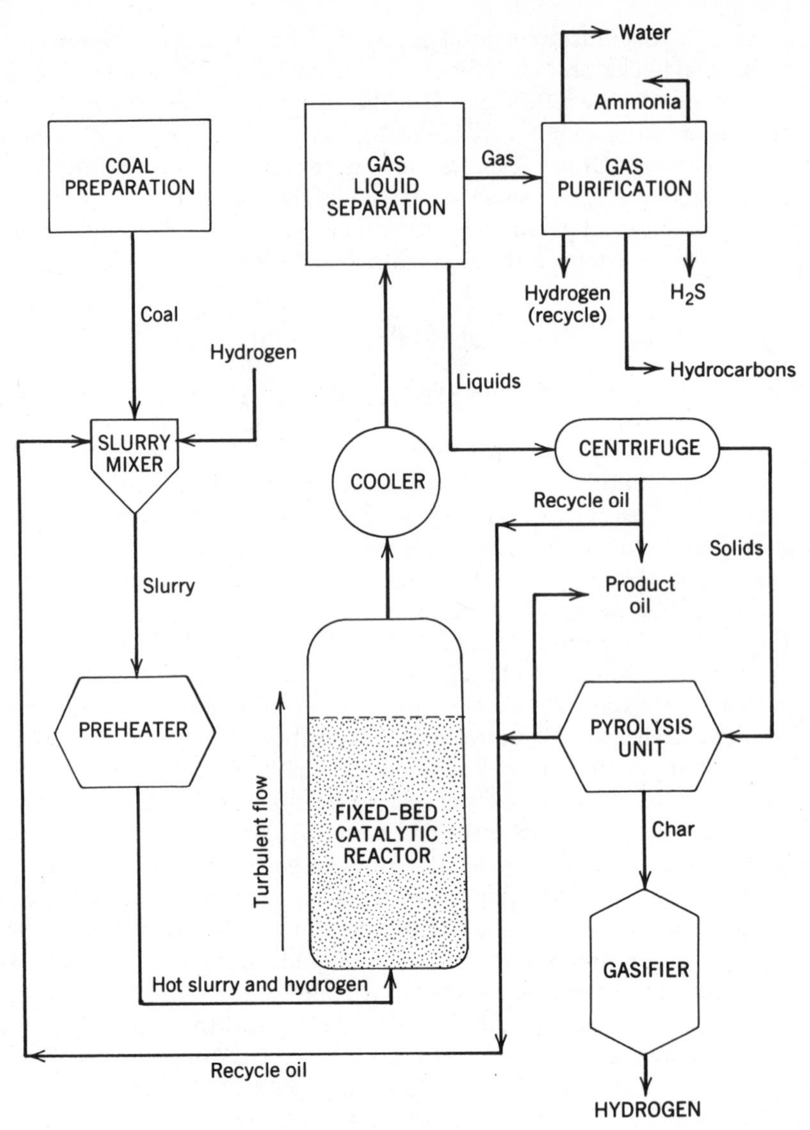

FIGURE 4.7 The SYNTHOIL process.

4.8.2 The H-Coal Process

The H-coal process was developed by Hydrocarbon Research, Inc., (HRI) under the joint sponsorship of the U.S. Department of Energy, the Commonwealth of Kentucky, and an industrial consortium headed by Ashland Synthetic Fuels. Like the SYNTHOIL process, the H-coal process is a catalytic hydroliquefaction process (55,56). It is flexible in that it can use a range of coals from lignite through bituminous and can be operated in the fuel oil mode or in the syncrude mode. The goal of the fuel oil mode is the production of a heavy oil suitable for use as a boiler fuel. The goal of the syncrude mode is the production of a synthetic crude oil that can be upgraded to produce transportation fuels. The H-coal process is illustrated in Figure 4.8.

The coal is crushed (through 60 mesh), dried, and slurried with a coal-derived recycle oil. The slurry is preheated to around 350° C, mixed with hydrogen, and pumped into the H-coal reactor. The reactor contains a catalyst bed of pelletized cobalt molybdate on alumina. The catalyst bed is ebullated, or fluidized, by the upward motion of the coal slurry and hydrogen (Fig. 4.9).

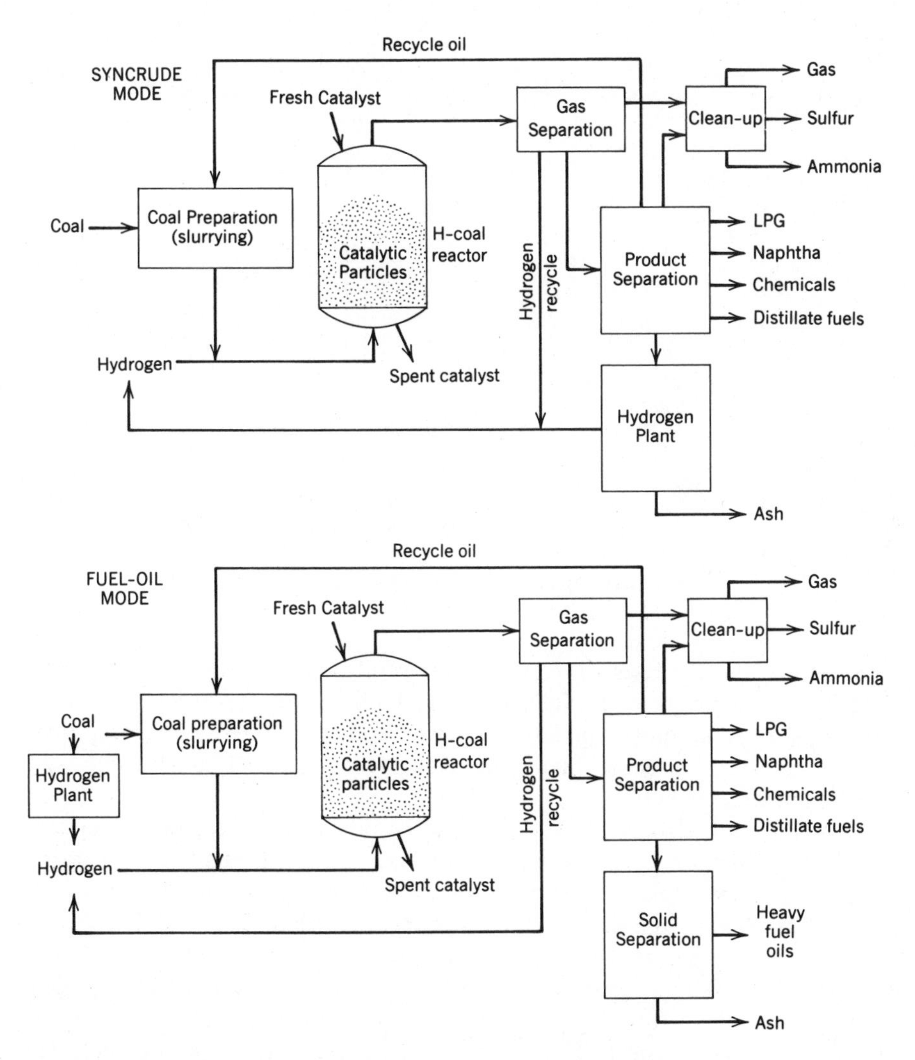

FIGURE 4.8 The H-coal process. (*Source:* Reprinted with permission from *H-Coal: A Future Energy Source,* Ashland Development Co.)

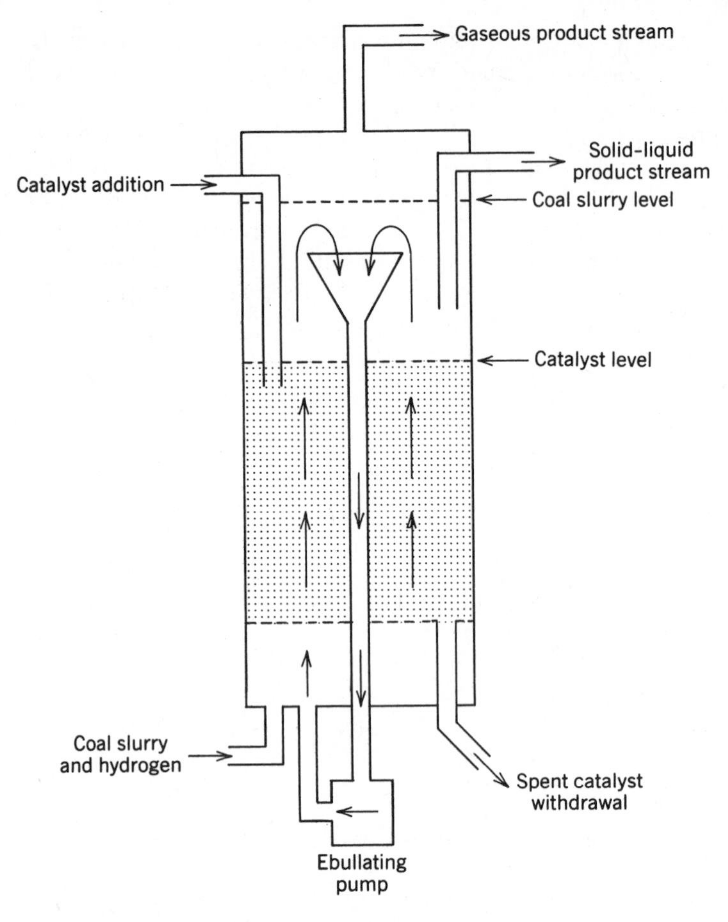

FIGURE 4.9 H-coal ebullated-bed reactor.

The ebullation increases contact with the catalyst and also facilitates the removal and replacement of catalyst. The reaction conditions inside the reactor are maintained at approximately 450° C and about 20–24 MPa hydrogen pressure. This is, of course, a direct catalytic approach to coal liquefaction using heterogeneous catalysis.

Placing the hydrogenation catalyst in the liquefaction reactor offers some advantages and some disadvantages. It is very likely that having the catalyst immediately available to renew the solvent and to hydrogenate the coal fragments being split from the coal matrix reduces the char formation that occurs when the fragments are allowed to couple in the absence of hydrogen. Direct catalytic hydroliquefaction also yields a lighter oil with somewhat lower heteroatom content. However, there are problems associated with this mode of operation. First is the lack of opportunity to separately optimize both the liquefaction step and the recycle solvent hydrogenation step of the process.

Second is the loss of catalytic activity as a result of the accumulation of mineral matter within the pore structure of the catalyst pellets. This necessitates frequent replacement of the catalyst and significantly increases the overall cost.

Vapor from the top of the H-coal reactor is cooled to separate the gases from the liquid products. The gases are scrubbed to remove ammonia, hydrogen sulfide, and hydrogen from the hydrocarbons. The hydrogen from the gas stream is recycled to the liquefaction reactor. The liquid is first subjected to atmospheric distillation to remove distillate products. The liquid stream from the reactor (containing mineral matter and unreacted coal) is fed into a flash separator. The distillate fraction undergoes atmospheric distillation, while the nonvolatile component is fed to a hydroclone. The overflow from the hydroclone is used for recycle oil, while the underflow is subjected to demineralizing (fuel oil mode) or vacuum distillation (syncrude mode). The vacuum still bottoms and unreacted coal may be treated as a waste or subjected to gasification to produce hydrogen for the process.

Operation in the syncrude mode requires more extensive hydrogenation. This is generally accomplished by longer residence time in the liquefaction

FIGURE 4.10 H-coal pilot plant. (*Source:* Reprinted with permission from *H-Coal Pilot Plant, Catlettsburg, Kentucky,* Ashland Development Co.)

reactor (about 1 h, as opposed to 0.5 h in the fuel oil mode). In the fuel oil mode the product is a low-volatile heavy oil with a higher heteroatom content. The overall product is approximately three barrels of oil per ton of bituminous coal.

The pilot plant at Catlettsburg, Kentucky, was capable of processing between 200 and 530 tons of coal per day (depending on the mode of operation) (Fig. 4.10). This was the largest coal liquefaction plant to be operated in the United States and was considered a technical success. The 18,000 ton per day Breckinridge Project (utilizing H-coal technology) was scheduled for construction between 1983 and 1989. Funding problems, however, resulted in the cancellation of the project.

4.8.3 The Conoco Zinc Chloride Process

The zinc chloride process was developed by Continental Oil Company with support from the Shell Development Company and the U.S. Department of Energy. This coal liquefaction scheme, although considered a direct catalytic hydroliquefaction, is quite different from those previously presented. Zinc chloride is a very good Lewis-acid–type catalyst that shows catalytic activity for most Friedel-Crafts reactions. A Lewis acid is defined as a substance that can accept an electron pair to form a covalent bond. Zinc chloride is also a good cracking catalyst, especially when used in high concentrations (40). It is a superior catalyst for the hydrocracking of polynuclear aromatic compounds but shows relatively little activity toward hydrogenation or cracking of mononuclear aromatic compounds (41). In the liquefaction of coal, this should be most useful in reducing the number of polynuclear aromatic compounds in the coal-derived liquid. When zinc chloride is used in amounts comparable to the weight of the coal (1:1 catalyst-to-coal ratio), the hydrocracking reaction rapidly yields a liquid product with a higher percentage of oils in the gasoline or naphtha range and a lesser amount of preasphaltenes and asphaltenes than do other coal liquefaction schemes developed to date. The specific goal of the Conoco zinc chloride process is the catalytic hydrocracking of coal into a light oil suitable for further refining into a high-octane transportation fuel. A flowsheet for the overall process is shown in Figure 4.11.

The coal is crushed, dried, and slurried with a coal-derived recycle oil. The slurry is pumped into a liquefaction reactor, where it reacts with hydrogen and molten zinc chloride. The reactor is maintained at temperatures of 350–440° C at hydrogen pressures of 10–30 MPa. The vapors from the reactor are cooled and neutralized, and the gases are separated from liquid products. The hydrogen is removed from the hydrocarbons in the gas stream and recycled. The liquid product is first separated into a distillate fraction and a nondistillate, or recycle oil, fraction. The distillate fraction is then fractionated into naphtha, fuel oil, and recycle oil.

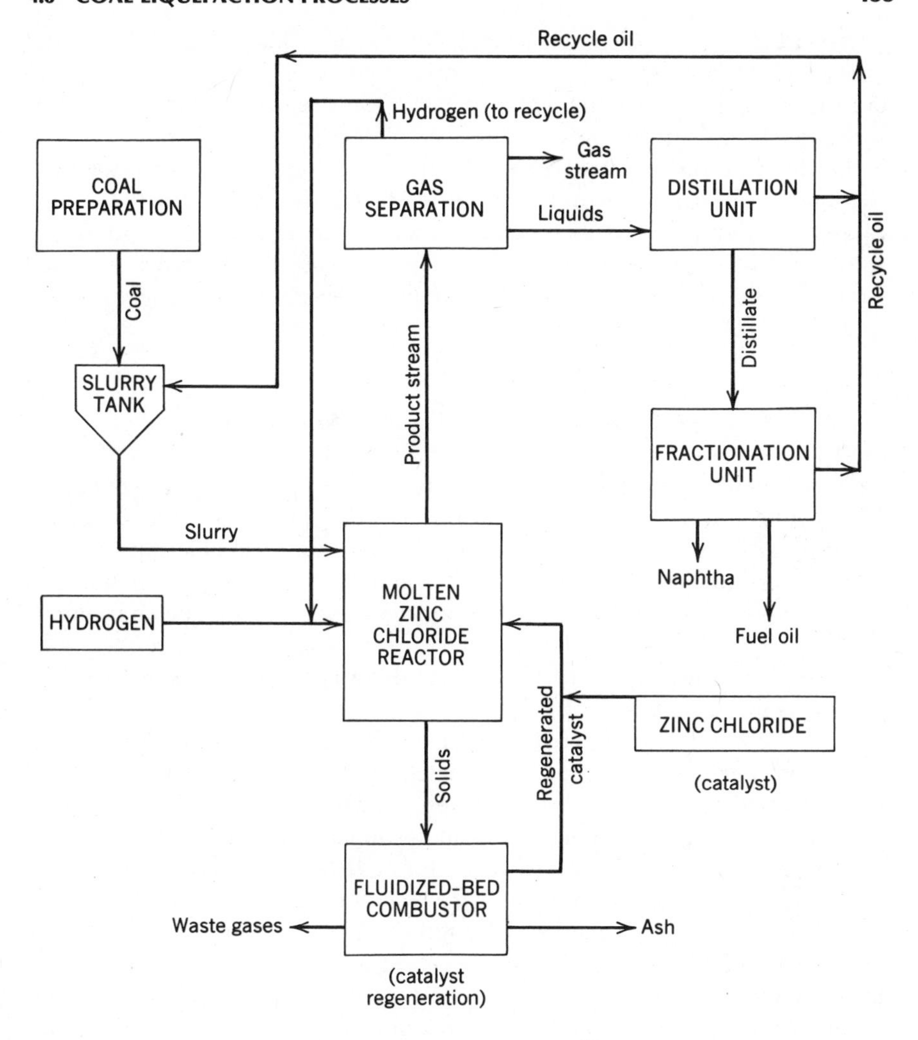

FIGURE 4.11 The Conoco zinc chloride process.

The oil from the zinc chloride process, unlike that from other liquefaction schemes, is very low in heteroatom content and easily refined into a transportation fuel. The heteroatoms present in the coal react with the zinc chloride catalyst. While this results in a much lower heteroatom content in the coal-derived oil, it also necessitates a regeneration of the zinc chloride catalyst:

$$ZnCl_2 + H_2S \longrightarrow ZnS + 2HCl$$

$$ZnCl_2 + H_2O \longrightarrow ZnO + 2HCl$$

$$ZnCl_2 + xNH_3 \longrightarrow ZnCl_2 \cdot xNH_3$$

$$ZnCl_2 \cdot xNH_3 + xHCl \longrightarrow ZnCl_2 \cdot xNH_4Cl$$

The spent catalyst is regenerated in a fluidized-bed oxidation unit. The oxidation process removes the carbon, nitrogen, and sulfur impurities and simultaneously vaporizes the zinc chloride (44). The regenerated zinc chloride is recycled into the liquefaction reactor.

The major problems associated with the zinc chloride process are the massive amounts of catalyst required, corrosion problems due to the acidity of the catalyst and by-products, and the necessity for the efficient recovery of the spent catalyst. The Conoco zinc chloride process has been operated in a 1 ton per day process development unit to evaluate the engineering problems and to study the economics of the process.

4.8.4 The Dow Coal Liquefaction Process

Another unique approach to direct catalytic hydroliquefaction of coal is the Dow coal liquefaction process, developed by the Dow Chemical Company (57,58). The Dow process employs a proprietary, water-soluble salt (or salts) as an expendable catalyst. The catalyst is added to the slurry as an emulsion in a process-derived oil. The flowsheet for the Dow process is shown in Figure 4.12.

The coal is crushed, dried, and mixed with a coal-derived recycle oil. The catalyst is prepared by dissolving the metal salt (or salts) in water and emulsifying the solution in coal-derived recycle oil. The emulsified catalyst is added to the coal slurry, containing approximately 40% coal by weight, and pumped into a preheater, where it is heated to approximately 420° C under a hydrogen pressure of about 14 MPa. The coal-catalyst-oil slurry is then pumped into a tubular, entrained-flow liquefaction reactor at 450–460° C and 14 MPa hydrogen pressure. The product stream for the reactor enters a gas-liquid separation unit. The gases are separated and are ultimately cleaned to recycle hydrogen and a hydrocarbon stream. The liquid stream is fed into hydroclones. The hydroclones employed return the overflow as recycle oil, while the underflow is fed into a deasphalter unit. The hydroclone does not efficiently remove the catalyst, and the recycle oil from the hydroclone contains catalyst, which is recycled.

The deasphalter unit employs a countercurrent liquid-liquid extraction process. A largely nonpolar deasphalter solvent is introduced, which causes the more polar asphaltenes to separate out as a second liquid phase. The oils in the product stream are much more soluble in the nonpolar phase than in the polar asphaltene phase. The asphaltene phase also agglomerates the solids present in the product stream. A viscous liquid containing the asphaltenes, mineral matter, and unreacted coal is removed from the bottom of the deasphalter. The top stream is flashed to recover the deasphalter solvent and a light oil product. The oil is split, with a portion going to the recycle stream and the majority to a distillation unit. The light oil is distilled to yield a naphtha fraction, a light fuel oil fraction, and a heavy fuel oil fraction. The viscous

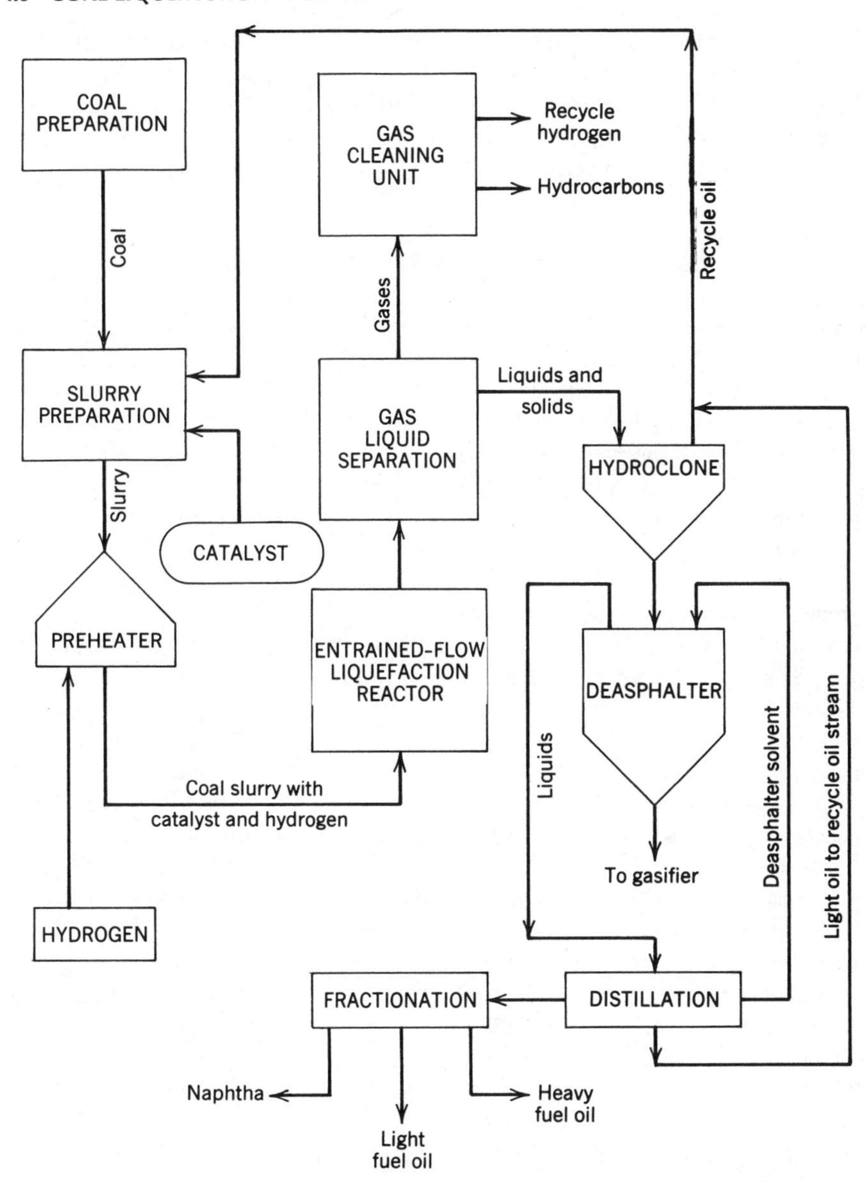

FIGURE 4.12 The Dow coal liquefaction process.

liquid from the deasphalter may be employed as a feedstock for a gasifier to prepare process hydrogen.

The Dow coal liquefaction process has been successfully operated on a 200 pound per day scale. This unique approach to coal liquefaction employs a very simple reactor design and emulsion technology to yield a catalyst that is both active and expendable. A number of other direct hydroliquefaction processes have been tested. The schemes discussed here represent an overview of a number of different approaches to direct hydroliquefaction of coal in the presence of an added catalyst. We shall now turn our attention to those coal liquefaction schemes that do not employ a catalyst (other than mineral matter) in the liquefaction reactor.

4.8.5 The Gulf Solvent-Refined Coal Process

A number of coal liquefaction schemes employ a noncatalytic direct hydroliquefaction step. The Gulf solvent-refined coal process (SRC) is one such scheme (59,60). The solvent-refined coal process was developed from German work (and work by Spencer Chemical Company) by Pittsburgh & Midway Coal Mining Company (a subsidiary of Gulf Oil Corporation) under the sponsorship of the Electric Power Research Institute (EPRI) and the U.S. Department of Energy. A simplified flowsheet for the SRC process is shown in Figure 4.13.

Solvent refining of coal was first accomplished on a large scale by Pott and Broche in Germany before World War II (61). The plant in Welheim, Germany, had a peak capacity of approximately 135 tons per day. The coal was slurried in a solvent (derived from coal tar) and heated under a pressure of 10–15 MPa at 430°C for 1 h. The slurry was cooled and filtered to yield a solvent-refined coal. This SRC product was used as a feedstock in the Bergius process for conversion into transportation fuels for the German war effort (1). A similar process was developed by F. Ude and T. W. Pfirrmann (62). This process treated the coal slurry with hydrogen at 410°C and 30 MPa.

The original Gulf solvent-refined coal process, now designated SRC-I, was designed to produce a low-sulfur and low-mineral matter solid fuel for use in electric power plants. This basic process was modified to produce a liquid rather than a solid product, and the new process was designated SRC-II. In the SRC-II process, the dried and pulverized coal (75% through 200 mesh) is slurried with a coal-derived solvent, mixed with hydrogen, and pumped through a slurry preheater. The preheater is maintained at 400–450°C and 10–14 MPa. The hot slurry then enters a vertical, concurrent upward, entrained-flow reactor called a dissolver. The reactor is maintained at around 440–470°C (pressure at 20–14 MPa). The effluent from the dissolver (liquefaction reactor) goes to a hot high-pressure separator. The top vapors are cooled, and the condensed liquid phase goes to the fractionating system, while the gas goes to an acid gas removal unit, which removes the CO_2 and H_2S. The treated gas then enters a cryogenic unit, where the methane and other light hydrocarbons are removed. The remainder of the gas (approximately 90% hydrogen) is mixed with additional hydrogen from a gasifier unit and is recycled to the preheater.

The slurry (liquid) from the hot high-pressure separator is split into two streams. One of these streams functions as the recycle oil or solvent and is

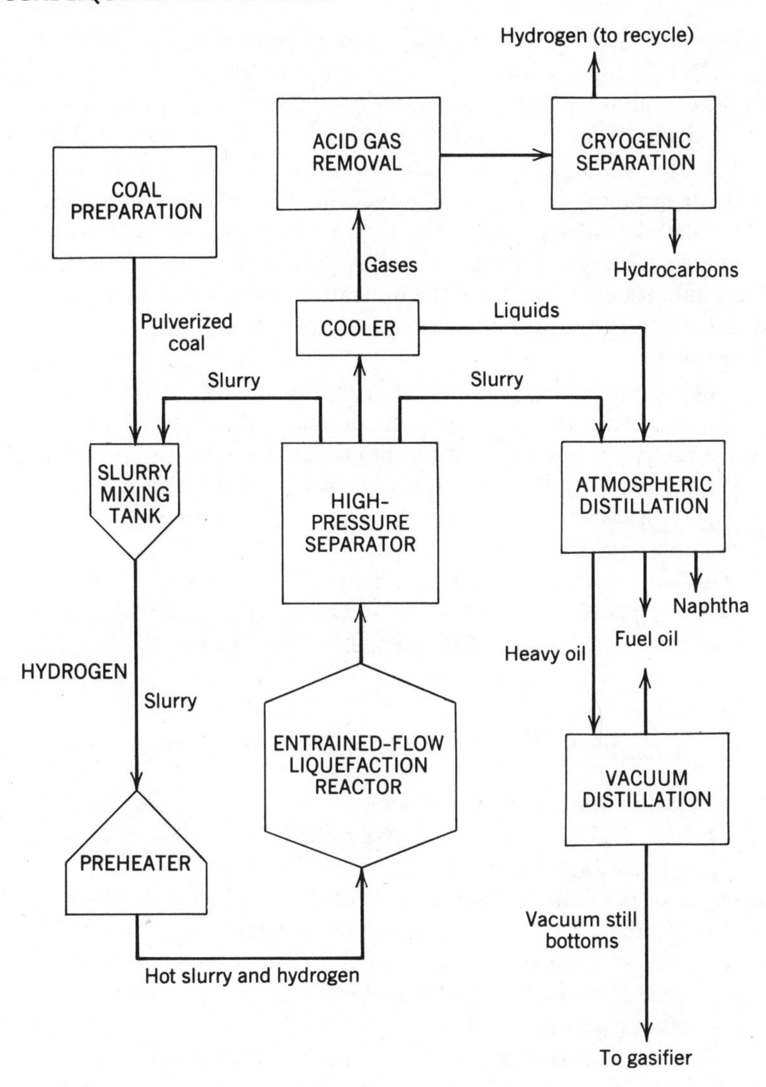

FIGURE 4.13 The Gulf solvent refined coal II process.

returned for the addition of more coal and then returned to the preheater. The second stream goes to the fractionating system for recovery of the light distillate (naphtha) and fuel oil. The nonvolatile material is then vacuum distilled to obtain additional fuel oil. The residue from the vacuum tower (vacuum still bottoms) is used as a feedstock for the gasifier unit to produce process hydrogen.

The more extensive hydrogenation required for the SRC-II process to produce liquid products is accomplished in several ways. First, the per-pass residence time in the dissolver (liquefaction reactor) is longer (1.0 h for SRC-II versus 0.4 h for SRC-I), and the hydrogen pressure is higher (roughly 13 versus 12 MPa). Second, the recycle of the product slurry returns much of the preasphaltene and asphaltene material to the reactor, so the total residence time is also significantly increased. Third, there is a greater concentration of mineral matter (catalyst) in the liquefaction reactor, which also aids in the overall hydrogenation process. The recycle feature for the high-boiling or nonvolatile material results in an effective increase of the severity of the process and thus results in a better oil yield. The adjustment of the per-pass residence time and the total residence time, and the utilization of the natural catalytic activity of the mineral matter result in an effective conversion of coal into gaseous and liquid products.

Some of the problems encountered in the SRC-II process include the limitations on the mineral matter content of the coal feedstock (the feed coal must have relatively low mineral matter content in order to prevent excessive buildup in the reactor), the viscosity of the slurry, and the solid content of the slurry. The high viscosity and relatively high-solids loading of the recycle solvent result in a very short pump life.

Two Gulf solvent-refined coal pilot plants were built and operated. A 5 ton per day SRC-I plant was constructed at Wilsonville, Alabama, and a 45 ton per day SRC-I/SRC-II plant was operated at Fort Lewis, Washington.

4.8.6 The Exxon Donor Solvent Process

The Exxon donor solvent process (EDS) was developed under the joint sponsorship of the U.S. Department of Energy, the Electric Power Research Institute, and an industrial consortium headed by Carter Oil Company, an affiliate of Exxon. Like the Gulf SRC-II process, the EDS process does not employ a catalyst (other than mineral matter) in the liquefaction reactor (63,64). The EDS process does, however, utilize a separate primary coal hydroliquefaction step and a secondary catalytic hydroprocessing, or solvent renewal, step. Such separation allows for the optimization of both steps in the coal conversion process. This optimization is very important in that the EDS process appears to be quite sensitive to the donor capacity (or reactivity) of the recycle solvent. The separation of the solvent hydroprocessing step allows the recycle solvent to be adjusted by catalytic hydrogenation to a specific donor activity (solvent quality index) deemed optimal for the process. The separation of the catalyst (a commercial pelletized cobalt molybdate on alumina) from the primary liquefaction reactor allows for a more flexible control of the extent of coal hydrogenation and has the additional advantage of preventing the relatively rapid loss of catalytic activity due to the accumulation of mineral matter in the catalyst. This is, of course, an important consideration in the overall economics of the EDS process. A simplified flowsheet for the Exxon donor solvent process is shown in Figure 4.14.

In the EDS process, the dried, pulverized coal is slurried with recycle sol-

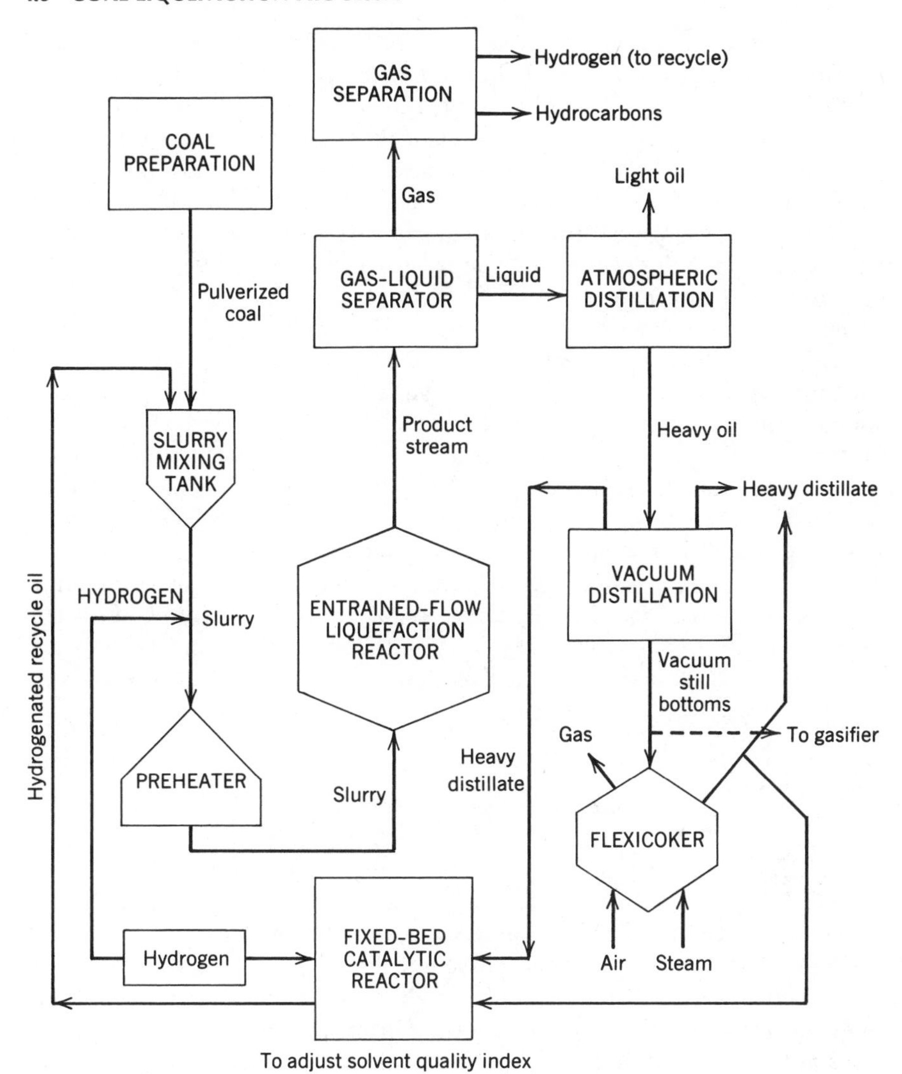

FIGURE 4.14 The Exxon donor solvent process.

vent of the desired solvent quality index, mixed with hydrogen, and pumped through a preheater into a vertical, entrained-flow reactor (the liquefaction reactor). The reactor is maintained at 425–480°C at 14–17 MPa hydrogen pressure. The residence time in the liquefaction reactor is from 0.5 to 0.75 h. After hydrogenation, the product goes to a gas separation unit, where hydrogen and hydrocarbons are separated from the liquid phase. The liquids are first subjected to atmospheric distillation to remove the light oils and finally to vacuum distillation to yield a heavy distillate. A portion of the distillate is taken as recycle oil and the remainder as product. The recycle solvent is hydrogenated in a fixed-bed catalytic reactor to the desired level of activity and recycled. The vacuum still bottoms are fed into an integrated coking (pyrolysis) and gasification reactor called a flexicoker. The flexicoker operates at approximately 450–650°C in the pyrolysis (coking) section and from 800–980°C in the gasifier section. Residence time ranges from 0.5 h to 1 h. The flexicoker provides additional liquid and gaseous product. As an alternative, the vacuum still bottoms may be fed to a standard gasifier unit to produce process hydrogen.

Some of the problems encountered in the EDS process include a relatively low yield of distillable oils (about 35% of the dry weight of the coal) and consequently a rather high yield of nonvolatile vacuum bottoms. This is a consequence of the noncatalytic nature of the conversion process. Since this is a rather mild hydroliquefaction, large amounts of preasphaltenes and asphaltenes remain unconverted into oils. In addition, the mineral matter in the feedstock must be carefully controlled to prevent excessive accumulation of mineral matter in the entrained-flow liquefaction reactor.

The Exxon donor solvent process has operated on a pilot plant scale second in size only to the H-coal process. A 225 ton per day pilot plant started operation in 1979 in Baytown, Texas. Overall liquid yield from EDS process averages around 2.6 barrels per ton of Illinois bituminous coal. Lower-rank coals produce somewhat lower yields.

4.8.7 The CO-STEAM Process

A unique coal liquefaction process in which there is no added catalyst is the CO-STEAM process. This liquefaction process was developed specificially for lignites by the Grand Forks Energy Research Center (now a part of the University of North Dakota) and the Pittsburgh Energy Technology Center. The CO-STEAM process uses carbon monoxide or synthesis gas ($CO + H_2$) and water (steam) in place of hydrogen for the conversion process (65–67). A catalyst is not required, since the low-rank coals tend to be very reactive and the mineral matter in these coals seems to have sufficient catalytic activity. The conversion of coal to heavy fuel oil, water, and gas has been reported to be approximately 90% at 450°C and 20.7 MPa for a lignite coal (68). For the lignites tested, the conversion yields in the noncatalyzed CO-STEAM process were higher than those obtained in the noncatalytic direct hydrogenation process. However, the CO-STEAM process is not as successful for the less reactive subbituminous and bituminous coals. It has been suggested that the superiority of the CO-STEAM process over direct hydrogenation for lignite may be attributed to the catalytic effect of alkali and alkaline earth metals present in these low-rank coals (66). These compounds are known to be effective catalysts in the carbon-steam and CO-steam reactions (69).

The liquid product from the CO-STEAM process contains few light oils.

The major product is a heavy viscous oil that may be suitable for use as a boiler fuel. Tests were conducted using both stirred-autoclave and tubular-flow reactors. Most of the tests have been done on a 120-lb/da or smaller scale.

The final technique for the liquefaction of coal to be discussed here is the only one currently being used on a relative large commercial scale (in the SASOL Project in South Africa). This is an indirect coal liquefaction process based on Fischer-Tropsch chemistry (Section 4.9.4). In this process, coal is first gasified to produce synthesis gas. The synthesis gas may then be "shifted" to adjust the H_2:CO ratio to the desired value, desulfurized, and then catalytically reacted to form a hydrocarbon mixture (fuel). Before discussing indirect coal liquefaction, we need to discuss briefly the coal gasification process.

4.9 COAL GASIFICATION

As an alternative to the production of liquid fuels from coal by pyrolysis and direct hydrogenation, both liquid and gaseous fuels can be prepared by the technique of coal gasification. The synthesis of liquid fuels via gasification may be based on Fischer-Tropsch chemistry or on the conversion of methanol (made from coal via gasification) to gasoline (Mobil M process). Gasification may also be used to produce synthesis gas ($CO + H_2$) for the manufacture of chemicals or for the production of gaseous fuels. The production of gaseous fuels may be divided into three rather broad classes based on the heating value of the fuel. Gas heating values are determined on the basis of the number of Btu per standard cubic foot (scf) of gas or the number of megajoules (MJ) per cubic meter of gas. The first class is called low-Btu gas and has a heating value of 100–200 Btu/scf (4–8 MJ/m^3). The second class is called medium-Btu or intermediate-Btu gas and has a heating value in the range of 200–500 Btu/scf (8–20 MJ/m^3). The third and final class is called high-Btu gas or substitute natural gas (SNG) (also known as synthetic natural gas) and must have a heating value above 900 Btu/scf (36 MJ/m^3). As a general rule, the low-Btu gas is the product of coal gasification using steam and air. Medium- or intermediate-Btu gas is the product of coal gasification using steam and oxygen (as opposed to air). The nitrogen from air remains in the low-Btu gas, acting as a diluent and reducing its heating value. The synthesis of high-Btu gas requires an upgrading of medium-Btu gas by such techniques as catalytic conversion and methanation. All three classes of fuel gases must be cleaned before use to remove particulates, tars, H_2S and CO_2.

One of the recent objectives in the gasification of coal is the preparation of pipeline-quality high-Btu or substitute natural gas. This objective may be achieved in four basic steps, as illustrated in Figure 4.15. The first step is pyrolysis, or devolatilization, of the coal. The second step involves the reaction of the devolatilized coal (char) with steam and either air or oxygen to produce a mixture of carbon monoxide and hydrogen called synthesis gas. In

Step 1. Pyrolysis, or devolatilization

$$\text{Coal} \xrightarrow[590\text{–}820^\circ\text{C}]{\text{heat}} CH_4 + \text{Char}$$

Step 2. Gasification, or the steam-carbon reaction

$$\underset{\text{Char}}{C} + \underset{\text{Steam}}{H_2O} \xrightarrow{927^\circ\text{C}} CO + H_2 \qquad \Delta H = 31.4 \text{ kcal/mole}$$

Step 3. Water-gas shift, or shift reaction

$$CO + H_2O \underset{}{\overset{327^\circ}{\rightleftharpoons}} H_2 + CO_2 \qquad \Delta H = -9.8 \text{ kcal/mole}$$

Step 4. Methane synthesis

a. Direct hydrogasification

$$C + 2H_2 \overset{927^\circ\text{C}}{\rightleftharpoons} CH_4 \qquad \Delta H = -17.9 \text{ kcal/mole}$$

b. Catalytic methanation

$$CO + 3H_2 \underset{\text{Ni}}{\overset{371^\circ\text{C}}{\rightleftharpoons}} CH_4 + H_2O \qquad \Delta H = -49.3 \text{ kcal/mole}$$

FIGURE 4.15 The basic coal gasification reactions.

the third step, some of the CO reacts with steam to produce more H_2. This reaction, used to adjust the H_2:CO ratio to the desired value, is called the water-gas shift or, simply, the shift reaction. Finally, in the fourth step the methane level is raised to the desired value by the synthesis of additional methane by either catalytic methanation or direct hydrogasification.

In the production of high-Btu gas, it is highly desirable to maximize the yield of methane from the pyrolysis step. Two ways in which the yield of methane may be enhanced are through the use of Lewis acid catalysts and rapid heating rates (43,70). After pyrolysis, rather large quantities of char remain. The production of additional methane from the residual char requires a supplemental hydrogen source. To produce this additional methane, a combination of steps 2, 3 and 4 is used.

The highly endothermic nature of the carbon-steam reaction (step 2) accounts for the large energy demand for the gasification process and thus a large portion of the process cost. Since step 2 is highly endothermic (ΔH = 31.4 kcal/mole), large amounts of heat must be supplied to the gasification reactor. This heat is most commonly supplied by the introduction of air or oxygen into the reactor so that the partial combustion of the char can be used as a heat source:

$$\underset{\text{char}}{C} + O_2 \longrightarrow CO_2 \qquad \Delta H = -94.1 \text{ kcal/mole}$$

Oxygen, while much more expensive than air to use, is desirable to produce a higher-Btu product gas. The presence of nitrogen as a diluent in air-blown gasifiers leads to a low-Btu product gas. Heat may also be added to the gasifier electrically or by the use of carriers, such as hot char or ash derived from char combustion, and recycled back into the gasifier.

The carbon dioxide produced in the partial combustion step can be reduced back to carbon monoxide by reaction with hot char in the reactor (Boudouard reaction):

$$\underset{\text{char}}{C} + CO_2 \longrightarrow 2CO \quad \Delta H = 41.3 \text{ kcal/mole}$$

In an efficient gasifier, there are both a relatively high conversion to CO and H_2 (step 2) and, likewise, a subsequent conversion of CO_2 to CO. The synthesis gas produced in the carbon-steam reaction may itself be burned as fuel or further converted to methane.

The water-gas shift, or shift reaction (step 3), may be used in either of two ways. First, it can be used to generate additional hydrogen within the gasifier itself, which helps promote the direct hydrogasification reaction (step 4a). The formation of methane via this route also helps provide heat for the carbon-steam reaction. Second, the shift reaction can be used externally from the gasifier to adjust the H_2:CO ratio to a value of 3 for the catalytic methanation reaction (step 4b). This reaction is highly exothermic, but the heat generated cannot be effectively utilized for the gasifier, since the reaction is carried out in a separate reactor and at a lower temperature (to favor methane formation) than is the highly endothermic carbon-steam reaction.

Finally, there are two basic ways in which the methane level can be boosted to produce high-Btu gas from coal. First is the direct hydrogenation of the char (step 4a). This reaction is accomplished directly in the gasifier and does not, therefore, require a separate reactor for the shift reaction. Second is the catalytic methanation process (step 4b). Catalytic methanation requires a separate reactor but occurs at lower temperature and is the easier process to operate efficiently.

Three basic problems are associated with the conversion of coal to high-Btu gas by direct hydrogasification. First, at 700–800° C, the temperature range that permits an efficient direct synthesis of methane, the rate of coal conversion is low. Increasing the reaction temperature above 800° C increases the rate of conversion, but the equilibrium content of methane in the gas stream decreases with increasing temperature. Gasifiers operating at high temperature produce little or no methane. Second, some kind of external hydrogen source is required to maintain the proper partial pressure required for efficient operation. Finally, the heat released in the hydrogenation step cannot be fully utilized.

The catalytic methanation reaction is carried out in a separate tube-wall reactor. The most effective methanation catalysts appear to be nickel and ruthenium, with nickel being the most widely used (71,72). The synthesis gas must be desulfurized before methanation, since sulfur quickly poisons the nickel catalysts employed. The reaction is typically carried out at 300–350° C at 2–3 MPa. Since the reaction is highly exothermic, provisions must be made to continuously withdraw heat from the methanation reactor.

One of the main problems in coal gasification is thermal efficiency. While the overall synthesis of methane is very nearly thermoneutral ($\Delta H = 3.7$ kcal/mole), the first step in the sequence is highly endothermic and requires a high reaction temperature:

Gasification	$2C + 2H_2O \rightleftharpoons 2CO + 2H_2$	$\Delta H = +62.8$ kcal
Shift	$CO + H_2O \rightleftharpoons CO_2 + H_2$	$\Delta H = -9.8$ kcal
Methanation	$3H_2 + CO \rightleftharpoons CH_4 + H_2O$	$\Delta H = -49.3$ kcal
Overall	$2C + 2H_2O \rightleftharpoons CH_4 + 2H_2$	$\Delta H = +3.7$ kcal

The last step is highly exothermic but requires a much lower reaction temperature. The heat evolved in the methanation step at 371° C cannot be recycled back to the gasifier, which must be maintained at a reaction temperature (gasification zone) of approximately 927° C. If one attempts to carry out the methanation at higher temperatures, the equilibrium becomes unfavorable.

As with coal liquefaction, coal (char) reactivity is an important concern. As a rule, one observes an increase in char reactivity toward gasification with decreasing rank of the coal from which the char is produced. A disadvantage of the gasification of bituminous and subbituminous coals is the formation of tars in the gas stream, which complicates the cleaning procedures. In addition, caking coals require pretreatment to prevent agglomeration in the reactor. The pretreatment most commonly involves a mild oxidation step and results in the loss of 5–20% of the weight of the coal. This, of course, adds significantly to the cost of the process.

The use of a catalyst to accelerate the steam gasification of coal would be highly desirable, since it would allow operation at higher coal throughputs and at a lower gasifier temperature. The most effective catalysts have been shown to be alkali metal salts, with the hydroxides and carbonates among the most active. It is interesting to note that the reactive low-rank coals (lignites) contain these kinds of compounds in their mineral matter. While in many cases the nature of the catalytic effect is not well understood, the catalytic activity of potassium carbonate has been ascribed to the following sequence of reactions (69):

$$K_2CO_3 + 2C \rightleftharpoons 2K + 3CO$$
$$2K + 2H_2O \rightleftharpoons 2KOH + H_2$$
$$CO + H_2O \rightleftharpoons CO_2 + H_2$$
$$2KOH + CO_2 \rightleftharpoons K_2CO_3 + H_2O$$
$$\overline{2C + 2H_2O \rightleftharpoons 2CO + 2H_2}$$

4.9.1 Coal Gasifiers

Four major types of gasifiers have been employed for coal gasification. They are classified in terms of the coal bed in the reactor and include moving-bed (or fixed-bed), fluidized-bed, entrained-bed (or entrained-flow), and molten-bath gasifiers. We shall consider the general operating characteristics of each type and then discuss briefly some representative commercial and experimental versions of each type.

A simplified schematic of a moving-bed gasifier is shown in Figure 4.16. In the moving-bed gasifier, the coal is dried and pulverized to sizes in the range of roughly 0.5–2 in. The pulverized coal is fed into the top of the reactor, where it falls by gravity toward the bed. As the coal falls, it is first dried, then devolatilized, and then gasified. Steam plus either air or oxygen is fed continuously into the bottom of the reactor to provide oxygen for the combustion of the char (which provides process heat) and steam for the carbon-steam and water-gas shift reactions. A temperature of 1,000–1,200° C is maintained in the

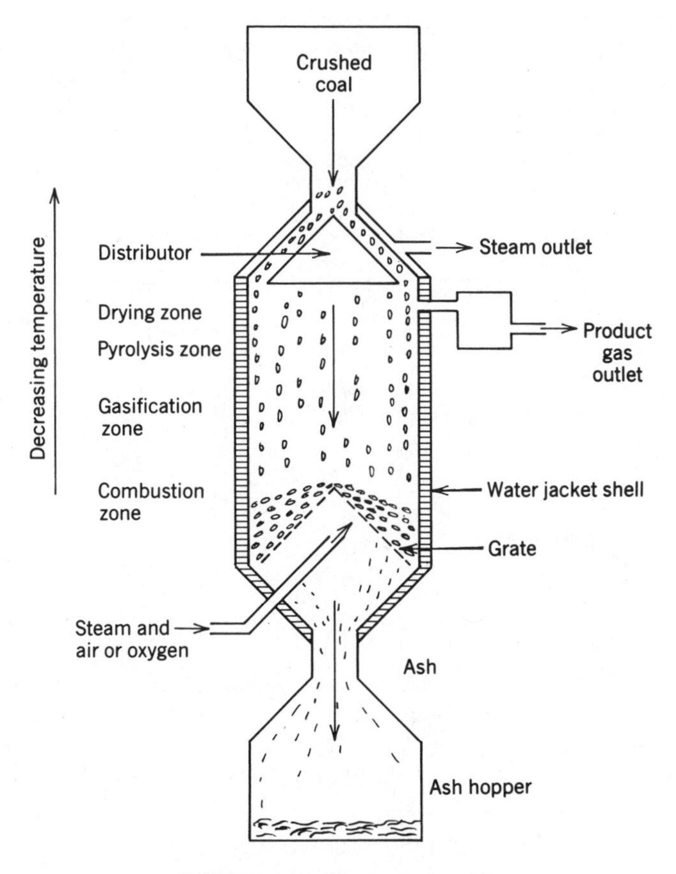

FIGURE 4.16 Moving-bed gasifier.

combustion zone. The temperature decreases progressively as materials move upward from the combustion zone through the gasification and devolatilization zones. The temperature of the effluent gas is in the range of 450–650° C, depending on feed coal and operating parameters.

Moving-bed gasifiers may be further classified as either one- or two-stage units, depending on whether there is a single outlet for the product gas or separate outlets for the devolatilizer (pyrolysis) gas and the product gas. The moving-bed reactor operates with a relatively high thermal efficiency and good carbon conversion. In addition, the effluent gas has a low degree of dust loading. The main disadvantages of the moving-bed system include the fact that it cannot tolerate a very high level of coal fines, and caking or agglomerating coals must undergo pretreatment.

The fluidized-bed gasifier is schematically illustrated in Figure 4.17. In the fluidized-bed design, the steam and air mixture flow upward through the unit at a rate that supports the coal particles in the gas stream. For fluidized-bed operation, the coal must be pulverized to about 10–20 mesh. The fluidized-bed reactor offers several advantages. This system provides an excellent mixing of the gases and the coal particles. The smaller particle size (larger surface area) also appears to give higher reaction rates than in moving-bed systems. In addition, the fluidized-bed reactor tolerates a wider variation in feed coal and

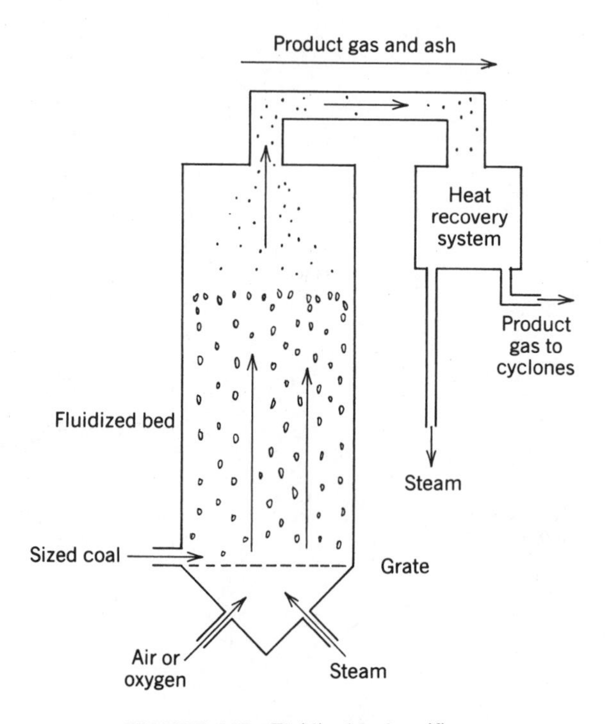

FIGURE 4.17 Fluidized-bed gasifier.

takes a higher rate of coal throughput. The disadvantages of the fluidized-bed gasifier are lower thermal efficiency, higher operating temperature, higher dust loading in the effluent gas, and shorter residence time in the reactor. The shorter residence times necessitates the use of a more reactive, lower-rank coal.

The entrained-flow gasifier is illustrated in Figure 4.18. In this design, the coal is first dried and pulverized to around 75% through 200 mesh and then fed, together with steam and either air or oxygen, into the bottom of the reactor. The very fine particles of coal are then carried upward along with the gas stream (or entrained flow). Because of the limited amount of contact between the coal particles in the gas stream, the caking, or agglomerating, problem is significantly reduced. Consequently, most coals can be used as feedstock without any pretreatment. The degree of coal conversion is limited, however, by the short residence time in the reactor. Higher reaction temperatures (1,600–1,950° C) are therefore required to overcome the low conversion yields associated with the short residence time in the reactor. The high-temperature operation also reduces the tar and methane levels in the effluent gas. The reduction in methane content may be desirable if the overall goal of the gasification is to produce synthesis gas and not fuel gas.

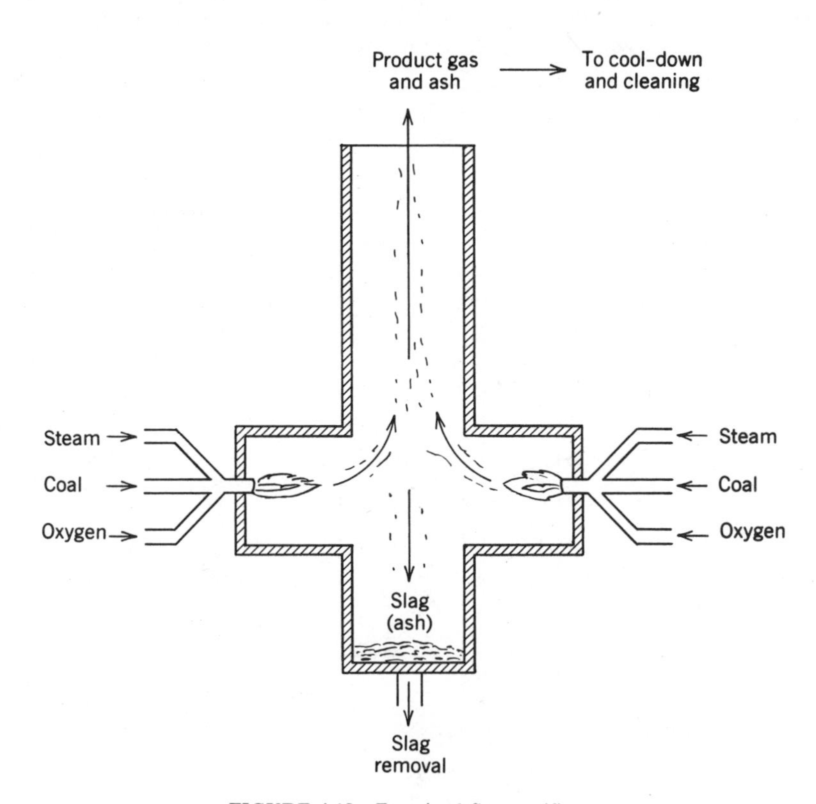

FIGURE 4.18 Entrained-flow gasifier.

Several systems have been designed to use the molten-bath technique (Fig. 4.19). The variety of molten baths employed include iron (ATGAS), sodium carbonate (Kellogg and Rockwell), and coal ash (Rummel slag bath). Pulverized coal is fed with oxygen and steam into the molten bath at temperatures from 900–1,400°C. Pretreatment of the coal is not required, and the ash can be removed from the top of the bath as slag. It is also possible to employ lime in the bath to trap sulfur and allow the gasification of high-sulfur coal in an environmentally acceptable manner.

4.9.2 Representative Coal Gasification Processes

A number of coal gasification processes are in various stages of development. These range from commercial to highly experimental. We shall consider here only a very brief outline of a few of the representative processes employing the different types of gasification reactors.

The oldest commercial coal gasification process employing a moving-bed reactor, the Lurgi process was first used commercially in Germany in 1936. It employs a medium-pressure gasification reactor operating at 2.6–3.1 MPa and a temperature of 620–760°C (73,74). In operation, crushed coal (⅛ in–2 in size range) falls through a rising stream of gases introduced through holes in a rotating grate at the bottom of the reactor. Immediately above the grate is a relatively narrow combustion zone, where the carbon remaining in the char is

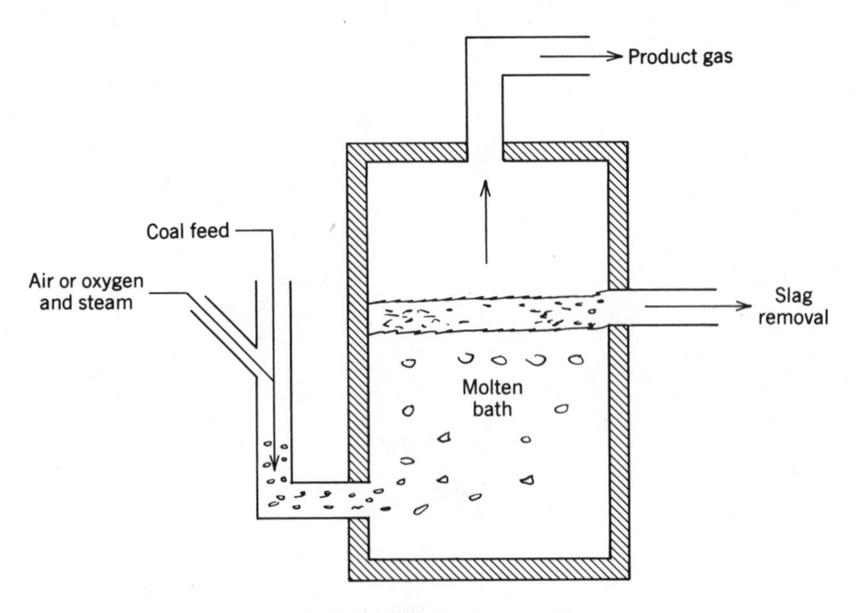

FIGURE 4.19 Molten-bath gasifier.

burned to provide process heat and CO_2. The hot gases rise through the bed of char, giving up heat to sustain the highly endothermic carbon-steam and Boudouard reactions. As the gases continue to rise and cool, the carbon-steam reaction stops, and the heat is used to devolatilize the coal. The effluent gases are then quenched with water and processed to remove tar, particulates, and acid gases (H_2S and CO_2).

The temperature in the combustion zone must be carefully controlled to keep it below the ash fusion temperature of the coal in use to prevent agglomeration (which would plug the grate). Large quantities of steam are injected to aid in temperature control. Much of this added steam is ultimately converted into hydrogen in the gasifier, but large amounts of water vapor are present in the effluent and must be removed. The heating value of the product gas is approximately 300 Btu/scf (12 MJ/m^3) when the reactor is oxygen blown and roughly 200 Btu/scf (8 MJ/m^3) when it is air blown.

The Wellman-Galusha process is a moving-bed gasification process that has been commercially available for almost 40 years. This process operates at atmospheric pressure and a temperature of 538–650°C (75). This gasifier can use most coals. The bed is stirred to reduce channeling and agglomeration. Likewise, the Woodall-Duckham/gas integrale system has been available since the 1940s (76). This system employs a two-stage gasifier. The top stage is vertical retort to carbonize the coal to obtain a pyrolysis gas. The lower stage gasifies the residual char to produce process heat and synthesis gas.

The Winkler process is the oldest commercial process employing fluidized-bed technology. This process was developed in Europe in the 1920s (77). Crushed coal (of roughly ⅜-in top size) is dried and fed into a fluidized-bed reactor by means of a variable-speed screw feeder. The gasifier operates at atmospheric pressure and a temperature of 815–1010°C. The high operating temperature leaves very little tar and hydrocarbons in the product gas stream. The gas stream, which may contain up to 70% of the ash, is cleaned by cyclones, water scrubbers, and electrostatic precipitators. The product gas is rich in CO and H_2 and has a heating value of roughly 275 Btu/scf (11 MJ/m^3).

A unique fluidized-bed process is the CO_2-acceptor process developed by the Conoco Coal Development Company (78,79). This process employs a fluidized-bed reactor at 1.0–1.4 MPa pressure and a temperature of 800–870°C. The lignite-derived char is significantly more reactive toward steam than is bituminous-derived char. Thus, the rate of the carbon-steam reaction for the bituminous-derived char is very slow at this operating temperature. The CO_2-acceptor process is not, therefore, useful for the higher-rank coals. The unique feature of the process is the manner in which some of the heat is derived for the carbon-steam reaction. The CO_2-acceptor process utilizes the strongly exothermic reaction between calcined dolomite ($MgO{\cdot}CaO$) and CO_2 to provide process heat and simultaneously remove CO_2 and H_2S from the gas stream:

Carbon-steam reaction

$$C + H_2O \longrightarrow CO + H_2 \qquad \Delta H = +31.4 \text{ kcal/mole}$$

Dolomite reaction

$$CaO + CO_2 \longrightarrow CaCO_3 \qquad \Delta H = -42.5 \text{ kcal/mole}$$

Desulfurization reaction

$$CaO + H_2S \longrightarrow CaS + H_2O \qquad \Delta H = -15.1 \text{ kcal/mole}$$

In operation, the crushed coal is preheated to 200–260°C and fed into the bottom of the char phase of the gasifier, where it is fluidized with steam. As the reaction progresses, the denser spent dolomite moves toward the bottom of the fluidized bed, where it is removed and sent along with char to the regenerator. The char and spent dolomite are burned with air in the regenerator at 1,065°C to calcine (regenerate) the dolomite before recycling:

Regeneration reaction

$$MgO{\cdot}CaCO_3 \longrightarrow MgO{\cdot}CaO + CO_2$$

Dolomite is fed into the top of the reactor, where it falls through the devolatilization zone, accepting CO_2 released by decarboxylation of the feed lignite. The CO_2 removal from the product gas stream not only provides process heat for the carbon-steam reaction but also favors formation of hydrogen via the shift reaction.

The effluent gas from the top of the reactor is quenched and cleaned to provide a product gas with an H_2:CO ratio of approximately 3 (ideal for methanation) and a heating value of roughly 440 Btu/scf (17.6 MJ/m^3). The CO_2-acceptor process offers several advantages. First, the process heat is supplied in situ without need for an external oxygen source. Second, a separate water-gas shift reactor is not required (the H_2:CO ratio is already almost ideal). Third, the cleanup operation is simplified by the low sulfur levels in the gas stream. There are also some disadvantages: only lignites are reactive enough to use in this process at normal operating temperatures, and supplying, regenerating, and recycling the dolomite entails additional costs. Conoco has operated the CO_2-acceptor process on a 40 ton per day scale at their pilot plant.

The synthane process is an experimental fluidized-bed design developed by the U.S. Bureau of Mines (now a part of the U.S. Department of Energy) (80). This process uses steam and oxygen to gasify coal in a fluidized bed at a pressure in the 4.1–6.9 MPa range and at a temperature of 816–982°C. The coal is dried, pulverized to 20–200 mesh, and fed into the fluidized-bed (steam-oxygen) pretreater stage at 427°C. The pretreater is used to prevent caking. The treated coal enters the gasifier stage, where it is fluidized with steam and oxygen. The integration of pretreatment and gasification is a key feature of the synthane process. The effluent gas is taken off the top of the reactor and the char-ash mixture from the bottom. The effluent gas contains a relatively high methane content and is a good candidate for conversion to a high-Btu fuel gas via catalytic methanation. Such a conversion requires removal of the tars, particulates, and sulfur followed by methanation. The methanation may be accomplished via tube-wall reactors in which Raney nickel is sprayed on the tube walls. The synthane process may use lignite, subbituminous, and bituminous coals. The process is still experimental and has yet to be demonstrated on a large scale.

The oldest commercial process using entrained-flow technology is the Koppers-Totzek process, developed in 1948 (81). In the Koppers-Totzek process, the coal is very finely pulverized and fed into the reactor with steam and oxygen. In contrast to the moving-bed and fluidized-bed processes, there are few limitations on the nature of the feed coal in terms of caking behavior and mineral matter properties. The original process operates at or near atmospheric pressure (0.1 MPa). More recently, experiments have been conducted using an entrained-flow Koppers-Totzek system at pressures of 1.5–3.0 MPa (82).

In operation, the Koppers-Totzek process feeds very finely pulverized coal along with steam and oxygen into a two- or four-burner, refractory-lined, cyclindrical reactor at 0.1 MPa pressure. Temperatures in the reactor are typically between 1,900°C and 2,000°C. Because of the very high temperature, the ash agglomerates and drops out of the combustion zone as a molten slag (which is withdrawn from the bottom of the reactor). The hot effluent gases are quenched and cleaned. The product gas contains almost no tars or hydrocarbons (methane) and is predominantly synthesis gas. The product has a heating value of approximately 280 Btu/scf (11.2 MJ/m^3) and is currently used for the synthesis of ammonia.

Texaco has modified its partial oxidation process, which has been used to gasify crude oil, to gasify coal (83). This process also employs entrained-flow technology. Pulverized coal (70% through 200 mesh) is slurried with water, preheated, and injected with oxygen into a gasifier operating at 1,100–1,370°C and a pressure of 2–8.5 MPa. The Texaco gasifier is a vertical, refractory-lined, cylindrical reactor with a gasification zone and a slag quench zone at the bottom. As in the Koppers-Totzek process, the effluent gas stream has little or no hydrocarbon content. The hydrocarbon products and tars are decomposed at the very high operating temperature. The Texaco process is used to manufacture synthesis gas.

The BI-GAS process, developed by Bituminous Coal Research, Inc., employs a two-stage entrained-flow reactor operating under high pressure (84). In operation, the dried, pulverized coal is fed into the top of the second stage of the reactor. This stage is kept at 927–982°C and 10.4 MPa pressure. The high pressure is employed in an attempt to maximize the yield of methane from the process. The fact that methane synthesis is an exothermic process reduces the external heat requirement and, thus, the oxygen consumption. The injected coal encounters a stream of hot synthesis gas, which was produced in the first stage of the reactor. The coal reacts with the hot gas to form methane and more synthesis gas. The char from the second stage is entrained in the product gas stream and removed with the effluent. The char is separated from the product gas and recycled back to the first stage of the gasifier, where it is gasified with oxygen and steam to produce both the synthesis gas and heat required for the second stage. One of the attractive features of the BI-GAS process is the higher than normal quantities of methane in the product gas. This process would be suitable for catalytic methanation to yield a high-Btu fuel gas.

The hydrane process is an experimental design of the U.S. Bureau of Mines and employs an entrained-flow and fluidized-bed reactor (85). The unique feature of this process is that essentially all of the product methane is reproduced in a single-stage reactor by direct hydrogenation or hydrogasification of the coal. This process operates at temperatures of 930–980°C and at a pressure of 6.9 MPa. Hydrogen for the process may be made by gasification of the char produced in the reactor. The dried, pulverized coal is injected into the top of the reactor, where it is hydrogenated as it falls toward the bottom of the reactor. The char then reacts with additional hydrogen in a fluidized-bed at the bottom of the reactor.

Finally, a number of coal gasification schemes employ a molten-bath reactor. With the exception of the Rummel slag bath process, these are experimental systems that have yet to be developed commercially. The ATGAS process (developed by Applied Technology Corporation) injects pulverized coal and lime into a bath of molten iron at 1,478°C (86). The coal is first devolatilized, leaving the fixed, or nonvolatile, carbon in the bath. The fixed carbon is then oxidized to CO by a stream of oxygen passed through the molten iron bath. Any sulfur present in the coal reacts with lime (on top of the bath) to produce calcium sulfide, which forms a slag along with the ash. The slag may then be skimmed off the top of the bath. The effluent gas is quenched and cleaned to produce a product with a heating value of approximately 460 Btu/scf (18.4 MJ/m^3).

Both the M. W. Kellogg Company and Rockwell International Corporation have developed molten-bath reactors that consist chiefly of molten sodium carbonate (87,88). The Kellogg molten salt process operates at 927°C and a pressure of 8.3 MPa. The pulverized coal is injected into the molten-bath reactor along with steam and oxygen. The coal is gasified by reaction with steam and oxygen in the bath. This process utilizes the catalytic effect of the alkali metal salts on the carbon-steam reaction. Continuous gasification and hot-gas cleanup occur simultaneously in the single reactor. Few impurities are present in the effluent gas stream. In addition, very little processing of the feed coal is required. The product gas is reported to have a heating value of approximately 330 Btu/scf (13.2 MJ/m^3).

The Rummel slag bath process employs a bath of molten ash (slag) for the gasification process. This commercial process was developed in Germany by Union Rheinische Braunkohlen Kraftstoff A. G. (89). The coal, along with steam and either air or oxygen, is injected directly into the molten slag. The contact between the coal and the molten slag results in efficient carbon conversion and yields a product gas with a heating value of about 270 Btu/scf (10.8 MJ/m^3).

4.9.3 Underground Coal Gasification

In addition to the coal gasification reactors previously described, there have been a variety of experiments, especially by the Soviet Union, on the in situ, or underground, gasification of coal (90,91). This technique offers the obvious advantages of being able to utilize coal reserves that would otherwise be very difficult to mine. Coal contained in steeply inclined seams and in seams with an excessively thick overburden are prime candidates for in situ processing.

A series of boreholes is drilled through the overburden into the coal seam. Since the normal permeability of the coal seam is not sufficient to permit gas movement through the coal seam, it is necessary to enhance the permeability by blasting to fracture the coal. This process is called linking the boreholes. The coal at the base of a borehole is ignited, and air and steam are injected. This borehole is called the injection well. The injection of air forces the combustion zone to move along the coal seam. Preceding the combustion zone are the gasification zone, the pyrolysis zone, and the drying zone. This is analogous to the operation of a moving-bed gasifier. As the gasification of the seam occurs, the gaseous products are withdrawn from the other boreholes (or production wells). The heating value of the product from an air-blown in situ gasifier is roughly 100 Btu/scf (4 MJ/m^3). The low heat content of the gas necessitates on-site utilization of the product gas.

Several problems are associated with underground coal gasification. First, there is little opportunity to control the quality of the product gas. The rate at which gas is produced is also difficult to control. As with other techniques of coal utilization, there are always environmental concerns, including land subsidence. More complete discussion of the techniques and problems associated with underground gasification is presented in excellent reviews by P. N. Thompson and by D. W. Gregg and T. F. Edgar (90,91).

4.9.4 Indirect Coal Liquefaction

In addition to the production of fuel gas, coal may be gasified to prepare synthesis gas ($CO + H_2$), which may be used to prepare a variety of products, including ammonia, methanol, and liquid hydrocarbon fuels. The synthesis of hydrocarbons from carbon monoxide and hydrogen (synthesis gas) is called the Fischer-Tropsch synthesis and is a procedure for the indirect liquefaction of coal (92). This liquefaction procedure is, at least in theory, the least efficient of those we have discussed, since it requires the breakdown of the coal matrix into single carbon species (CO) followed by a reassembly process (Fischer-Tropsch). Ironically, this process is the only coal liquefaction scheme currently in use on a relatively large commercial scale. The Republic of South Africa is currently using the Fischer-Tropsch process on a commercial scale in their SASOL complex (93). Germany produced roughly 156 million barrels of synthetic petroleum annually using the Fischer-Tropsch process during the Second World War (3).

The basic reactions involved in the Fischer-Tropsch synthesis can be generalized as follows:

$$C + H_2O \longrightarrow CO + H_2$$

$$nCO + 2nH_2 \longrightarrow C_nH_{2n} + nH_2O$$

$$nCO + (2n + 1)H_2 \longrightarrow C_nH_{2n+2} + nH_2O$$

The exothermic nature of the process and the decrease in the total gas volume in going from reactants to products suggest the most suitable experimental conditions to use in order to maximize product yields. The process should be favored by high pressure and relatively low reaction temperature. In practice, the Fischer-Tropsch reaction is carried out at temperatures of 200–350°C and at pressures of 0.5–4.1 MPa. The H_2:CO ratio is typically around 2.2:1.

Fused magnetite (iron oxide) catalysts are commonly used in the commercial Fischer-Tropsch synthesis and in the synthesis of ammonia. Magnesium oxide (MgO) is frequently added as a structural, or surface, promoter, and potassium oxide (or other alkali metal oxide) is often added as a chemical promoter (94,95). The structural promoter functions to provide a stable, high-area catalyst, while the chemical promoter alters the selectivity of the process. The effectiveness of the alkali metal oxide promoter increases with increasing basicity. Increasing the basicity of the catalyst shifts the selectivity of the reaction toward the heavier or longer-chain hydrocarbon products (96). By the proper choice of catalyst basicity and H_2:CO ratio, the product selectivity in the Fischer-Tropsch process can be adjusted to yield from 5–75% methane. Likewise, the proportion of hydrocarbons in the gasoline range (roughly C_6 to C_{12}) can be adjusted to produce 0–40% of the total hydrocarbon yield.

A number of problems are associated with the liquefaction of coal by the Fischer-Tropsch process. First and most obvious is the inherent inefficiency of any process that takes the molecules apart and then rebuilds them. Second, the hydrocarbon mixture from the process is largely aliphatic, with a preponderance of straight-chain hydrocarbons. The branched-chain alkanes are primarily mono- and dimethyl-substituted chains. There are few highly branched systems in the product. This results in a product (fuel) with a fairly low octane number (approximately 40), which must be further processed to yield a suitable fuel. Third, there is a need to develop better catalysts that exert even more control over the product distribution. Fourth, the synthesis gas must have a very low sulfur content, since the iron-based catalysts are very sensitive to sulfur poisoning by H_2S in the synthesis gas (97). Finally, the highly exothermic nature of the reaction causes rather rapid deterioration of the process catalyst and requires effective methods of heat removal from the reactor.

Methanol can also be readily made from synthesis gas (98):

$$CO + 2H_2 \longrightarrow CH_3OH$$

In practice, synthesis gas (along with recycle gas from the reactor) is compressed and pumped into a reactor called a methanol converter. The methanol converter is a pressure reactor containing tubes filled with a Cu-Zn catalyst. The reactor is maintained at 250–270°C and a pressure of 5–8 MPa. The conversion reaction is exothermic, and the heat must be removed from the system. The effluent gas from the converter goes through a heat exchanger to heat the feed gases and is condensed to yield a mixture of methanol and water. The methanol may be separated by distillation, or the reaction mixture may be used as feed for the Mobil M (or MTG) process.

Methanol can itself be used as a motor fuel. It has been used as a racing fuel for some time. Its use as a fuel does present some special problems. Methanol is toxic, hygroscopic, and (for a given volume of fuel) contains about half the energy content of gasoline. There are, therefore, good reasons to convert methanol to gasoline.

The Mobil M process was developed by Mobil Oil Corporation for the purpose of converting methanol directly into high-octane gasoline (93). In this process, methanol is first dehydrated to yield dimethyl ether, followed by further dehydration to yield a series of hydrocarbon products:

$$2CH_3{-}OH \longrightarrow CH_3{-}O{-}CH_3 + H_2O$$

$$nCH_3{-}O{-}CH_3 \longrightarrow {-}(CH_2{-}CH_2)_n{-} + nH_2O$$

The distribution of hydrocarbon products is controlled by the zeolite catalyst. The zeolite is a crystalline aluminosilicate with pores, or cavities. It is the size and geometry of the zeolite pores that control product distribution. Molecules with critical dimensions larger than the pore size cannot be made in the process. Both fixed-bed and fluidized-bed reactors have been employed in developing the Mobil M process. The reaction conditions are typically 330–400°C and a pressure of around 2.2 MPa. The yield of gasoline is reported to be high, and the octane rating has been reported to be above 90 (as opposed to about 40 from the Fischer-Tropsch process). This indirect coal liquefaction scheme looks very promising.

SUMMARY

A number of schemes are available for the conversion of coal into liquid and gaseous fuels. The principal methods of coal liquefaction include pyrolysis, direct hydrogenation, donor solvent hydrogenation, and indirect liquefaction using either the Fischer-Tropsch process or the Mobile M process. Basic research in both chemistry and engineering is geared toward the development of better coal conversion techniques and improvement of some of the older processes. The principal limitations at this time include very poor thermal efficiency, poor selectivity in the product formation, and high hydrogen consumption. The presence of heteroatoms in the coal-derived liquids is also a major problem. All of these factors, along with the necessity of using high-pressure reactors, cause coal liquefaction to be an extremely expensive undertaking. Improvements in any one of these problem areas could go far toward potential commercialization. The commercial exploitation of coal conversion is very unlikely as long as a relatively cheap source of natural petroleum is available. As petroleum resources become more limited, however, the abundant coal reserves will become more and more attractive as potential sources of chemical feedstocks as well as sources of liquid and gaseous fuels.

REFERENCES

1. E.E. Donath, "Hydrogenation of Coal and Tar," in H.H. Lowry, ed., *Chemistry of Coal Utilization,* supplementary vol., Wiley, New York, 1963, pp. 1041–1080.
2. H. Tropsch, *Chem. Ber., 68A,* 169 (1935).
3. A.N. Stranges, *J. Chem. Educ., 60,* 617 (1983).
4. J.F. Jones, "Project COED: Char-Oil-Energy Development," in *Symposium Papers: Clean Fuels from Coal,* Institute of Gas Technology, Chicago, 1973, pp. 383–402.
5. M.I. Greene, *Prepr., Div. Fuel Chem., Am. Chem. Soc., 19*(1), 215 (1974).
6. H.E. Jacobs, J.F. Jones, and R.T. Eddinger, *Ind. Eng. Chem. Process Des. Dev., 10*(4), 558 (1971).
7. F.L. McCray, N. McClintock, and R. Bloom, *Coal Processing Technology,* vol. 5, American Institute of Chemical Engineers, New York, 1979, pp. 156–165.
8. F.B. Carlson, M.T. Atwood, and L.H. Yardumian, *Chem. Eng. Progr., 69*(3), 50 (1973).
9. D.H. Cartez and C.J. DaDelfa, *Hydrocarbon Processing, 60*(2), 111 (1981).
10. S. Weller, E.L. Clark, and M.G. Pelipetz, *Ind. Eng. Chem., 42,* 334 (1950).
11. P. Nowacki, *Coal Liquefaction Processes,* Noyes Data, Park Ridge, NJ, 1979, pp. 71–72.
12. I. Howard-Smith and G.J. Werner, *Coal Conversion Technology,* Noyes Data, Park Ridge, NJ, 1976, p. 17.
13. A. Sass, *Chem. Eng. Progr., 70*(1), 72 (1974).
14. C. Oberg and A. Falk, *Coal Processing Technology,* vol. 6, American Institute of Chemical Engineers, New York, 1980, pp. 159–165.
15. G.P. Curran, R. Struck, and E. Gorin, *Ind. Eng. Chem. Process Des. Dev., 6*(2), 166 (1967).
16. S.R. Gun, J.K. Sama, P.B. Chowdhury, S.K. Mukherjee, and D.K. Mukherjee, *Fuel, 58,* 171 (1979).

17. Gun et al., p. 176.
18. L.W. Vernon, *Prepr., Div. Fuel Chem., Am. Chem. Soc., 24*(2), 143 (1979); *Fuel, 59,* 102 (1980).
19. R. Yoshida, Y. Maekawa, T. Ishii, and G. Takeya, *Fuel, 55,* 337 (1976).
20. Yoshida et al., p. 341.
21. F. Yoshimura, S. Mitsui, and Y. Fushizaki, *Kogyo Kagaku Zasshi, 65,* 377 (1962).
22. W.H. Wiser, G. Hill, and N. Kertamus, *Prepr., Div. Fuel Chem., Am. Chem. Soc., 10*(2), 121 (1966).
23. W.B. Davies and D.J. Brown, *Nature, 213*(5071), 64 (1967).
24. P. Hanbaba, H. Juentgen, and W. Peters, *Brennst. Chem., 49*(12), 368 (1968).
25. R.J. Pugmire, D.M. Grant, K.W. Zilm, L.L. Anderson, A.G. Oblad, and R.E. Wood, *Fuel, 56,* 295 (1977).
26. M. Anbar and G.A. St. John, *Fuel, 57,* 105 (1978).
27. D.M. Cantor, *Anal. Chem., 50,* 1185 (1978).
28. F.K. Schweighardt, H.L. Retcofsky, and R.A. Friedel, *Fuel, 55,* 313 (1976).
29. H.N. Sternberg, R. Raymond, and F.K. Schweighardt, *Science, 188,* 49 (1975).
30. R. Sexton, *Arch. Environ. Health, 1,* 181 (1960).
31. W.C. Hueper, *Ind. Med. Surgery, 25,* 51 (1956).
32. B.C. Gates, *Chemtech, 9*(2), 97–102 (1979).
33. W. Kawa, S. Friedman, W.R.K. Wu, L.V. Frank, and P.M. Yavorsky, *Prepr., Div. Fuel Chem., Am. Chem. Soc., 19*(1), 192 (1974).
34. T.E. Gangwer and H. Prasad, *Fuel, 58,* 577 (1979).
35. D.K. Mukherjee and P.B. Chowdhury, *Fuel, 55,* 4 (1976).
36. D. Gray, *Fuel, 57,* 213 (1978).
37. J.L. Cox, S.A. Wilcox, and G.L. Roberts, "Homogeneous Catalytic Hydrogenations of Complex Carbonaceous Substrates," in J.W. Larsen, Ed., "Organic Chemistry of Coal," *ACS Symp. Ser., 71,* American Chemical Society, Washington, 1978, pp. 186–214.
38. S. Friedman, S. Metlin, A. Svedi, and I. Wender, *J. Org. Chem., 24,* 1287 (1959).
39. N. Holy, T. Nalesnik, and S. McClanahan, *Fuel, 56,* 272 (1977).
40. C.W. Zielke, R.T. Struck, J.M. Evans, C.P. Costanza, and E. Gorin, *Ind. Eng. Chem. Process Des. Dev., 5*(2), 158 (1966).
41. Zielke et al., p. 151.
42. R.T. Struck, W.E. Clark, P.J. Dudt, W.A. Rosenhoover, C.W. Zielke, and E. Gorin, *Ind. Eng. Chem. Process Des. Dev., 8*(4), 546 (1969).
43. G.A. Mills, *Ind. Eng. Chem., 61*(7), 6 (1969).
44. C.W. Zielke, R.T. Struck, and E. Gorin, *Ind. Eng. Chem. Process Des. Dev., 8*(4), 552 (1969).
45. P.H. Given, D.C. Cronauer, W. Spackman, H.L. Lovel, A. Davis, and B. Biswas, *Fuel, 54,* 34 (1975).
46. J.F. Cudmore, *Fuel Processing Technology, 1,* 227 (1978).
47. D. Gray, G. Barrass, J. Jezko, and J.R. Kershaw, *Fuel, 59,* 146 (1980).
48. Given et al., p. 34.
49. Given et al., p. 40.
50. D.K. Mukhergee, J.K. Sama, P.B. Chowdhury, and A. Lahiri, "Effect of Reaction Variables on Hydrogenation of Baragolai (Assam) Coal," in *"Proceedings of the Symposium on Chemicals and Oil from Coal,"* Central Fuel Research Institute, Dhanbad, India, 1972, pp. 143–150.

51. M.B. Abdel-Baset, R.F. Yarzab, and P.H. Given, *Fuel, 57,* 89 (1978).
52. R.F. Yarzab, P.H. Given, W. Spackman, and A. Davis, *Fuel, 59,* 81 (1980).
53. S. Akhtar, N.J. Mazzocco, M. Weintraub, and P.M. Yavorsky, *Energy Communications, 1*(1), 21 (1975).
54. T. Derpich, S.H. Chanzg, and I.C. James, "Effect of Operating Variables on SYNTHOIL Reactor Performance," in T.N. Veziroglu, Ed., *"Proceedings and Condensed Papers from Miami International Conference on Alternative Energy Sources,"* University of Miami, Coral Gables, 1977, pp. 713–715.
55. H.H. Stotler and R.T. Schutter, *Coal Processing Technology,* vol. 5, 1979, pp. 73–77.
56. Nowacki, pp. 147–155.
57. N.G. Moll and G.J. Quarderer, *Chem. Eng. Progr., 75*(11), 46 (1979).
58. Nowacki, pp. 155–159.
59. B.K. Schmid, *Chem. Eng. Progr., 75*(4), 75 (1975).
60. B.K. Schmid and D.M. Jackson, *Proc. Pac. Chem. Eng. Congr., 2*(2), 908 (1977).
61. A. Pott and H. Broche, *Gluckauf, 69,* 903 (1933).
62. T.W. Pfirrmann, U.S. Patent 2,167,250 (1939).
63. W.N. Mitchell, K.L. Trachte, and S. Zaczepinsky, *Ind. Eng. Chem. Prod. Res. Dev., 18,* 311 (1979).
64. Nowacki, pp. 104–114.
65. H.R. Appell, I. Wender, and R.D. Miller, *Prepr., Div. Fuel Chem., Am. Chem. Soc., 12*(3), 220 (1968).
66. H.R. Appell, I. Wender, and R.D. Miller, *Prepr., Div. Fuel Chem., Am. Chem. Soc., 13*(4), 39 (1968).
67. H.R. Appell, I. Wender, and R.D. Miller, *Chem. Ind., 47,* 1703 (1969).
68. H.R. Appell, "COSTEAM—The Newest Coal Liquefaction Process," in F.J. Hendel, Ed., *"Alternate Energy Resources: Symposium Papers,"* Western Periodical Co., North Hollywood, 1976, pp. 74–79.
69. M.J. Veraa and A.T. Bell, *Fuel, 57,* 194 (1978).
70. R.A.A. Graff, S. Dobner, and A.M. Squires, *Fuel, 55* 109 (1976).
71. J.H. Field, J.J. Demeter, A.J. Forney, and D. Bienstock, *Ind. Eng. Chem. Prod. Des. Dev., 3*(2), 150 (1964).
72. E.R. Tucci and W.J. Thompson, *Hydrocarbon Processing, 58*(2), 123 (1979).
73. A. Verma, *Chemtech, 8*(6), 372–381 (1978).
74. A. Verma, *Chemtech, 8*(10), 626–638 (1978).
75. Howard-Smith and Werner, p. 57.
76. Howard-Smith and Werner, p. 64.
77. Howard-Smith and Werner, p. 58.
78. C.P. Curran, *Chem. Eng. Progr., 62*(2), 80 (1966).
79. C.P. Curran, C.E. Fink, and E. Gorin, *Ind. Eng. Chem. Process Des. Dev., 8*(4), 559 (1969).
80. A.J. Weiss, *Hydrocarbon Processing, 57*(6), 125 (1978).
81. H.J. Michaels and H.F. Leonard, *Chem. Eng. Progr., 74*(8), 85 (1978).
82. M.J. van der Burgt, *Hydrocarbon Processing, 58*(1), 161 (1979).
83. B. Cornils, J. Hibbel, P. Ruprecht, R. Durrfeld, and J. Langhoff, *Hydrocarbon Processing, 60*(1), 149 (1981).
84. W.P. Hegarty and B.E. Moody, *Chem. Eng. Progr., 69*(3), 37 (1973).
85. Howard-Smith and Werner, p. 47.
86. J.A. Karnavos, P.J. LaRosa, and E.A. Pelczarski, *Chem. Eng. Progr., 69*(3), 54 (1973).

87. A.E. Cover, W.C. Schreiner, and G.T. Skapendas, *Chem. Eng. Progr.*, 69(3), 31 (1973).
88. A.L. Koh, R.B. Harty, and J.G. Johnson, *Chem. Eng. Progr.*, *74*(8), 73 (1978).
89. Howard-Smith and Werner, p. 72.
90. P.N. Thompson, *Endeavor*, new series, *2*, 93 (1978).
91. D.W. Gregg and T.F. Edgar, *A.I.Ch.E.J.*, *24*, 753 (1978).
92. M.E. Dry, *Ind. Eng. Chem. Prod. Res. Dev.*, *15*(4), 282 (1976).
93. M. Singh, *Hydrocarbon Processing*, *60*(6), 138 (1981).
94. M.E. Dry, J.A.K. du Plessis, and G.M. Leuteritz, *J. Catalysis*, *6*, 194 (1966).
95. M.E. Dry and L.C. Ferreira, *J. Catalysis*, *7*, 352 (1967).
96. M.E. Dry and L.C. Ferreira, *J. Catalysis*, *11*, 18 (1968).
97. F.S. Karn, J.F. Shultz, R.E. Kelly, and R.B. Anderson, *Ind. Eng. Chem. Prod. Res. Dev.*, *2*(1), 43 (1963).
98. E.L. Muetterties and J. Stein, *Chem., Rev.*, *79*(6), 479 (1979).

5
THE ANALYTICAL CHEMISTRY OF COAL

Coal is a very heterogeneous material containing various combinations of organic matter and mineral matter. The principal elements in the organic matter are carbon, hydrogen, nitrogen, sulfur, and oxygen. The mineral matter may contain as many as 60 elements, which together make up the various minerals found in coal. These minerals include clay minerals, pyrite and/or marcasite, calcite, silica, and smaller amounts of other minerals. The analysis of coal, however, is determined from representative samples of the material and not from the individual components. Typical analyses for the principal ranks of coal are given in Table 5.1 (1).

Many of the methods of coal analysis are empirical and involve strict adherence to specified conditions, such as particle size, temperature, time and rate of heating, and so on. The establishment of specifications that are recognized as standards and supported by authoritative organizations is essential. The American National Standards Institute (ANSI) represents the United States at the international standards level and is similar to the British Standards Institute (BSI) in the United Kingdom. Unlike BSI, however, ANSI does not develop standards but looks to other organizations for such work. Committee D-5 of the American Society for Testing and Materials (ASTM) has the responsibility of developing standard procedures for coal analysis. This committee consists of approximately 300 members equally divided among producers, consumers, and those who have a general interest in coal.

The majority of the testing procedures developed by Committee D-5 that have international significance are also approved by ANSI. The use of ASTM procedures by coal testing laboratories is optional. However, these standards do have a certain degree of legal status and are used when coal is purchased according to a specification and penalty basis. Also, one of the criteria that one may use in judging the quality of an individual laboratory is the degree to which the laboratory is able to produce results that agree favorably with the

TABLE 5.1 Typical Composition and Physical Property Ranges for Various Ranks of Coal

	Anthracite	Bituminous	Subbituminous	Lignite
Moisture (%)	3–6	2–15	10–25	25–45
Volatile matter (%)	2–12	15–45	28–45	24–32
Fixed carbon (%)	75–85	50–70	30–57	25–30
Ash (%)	4–15	4–15	3–10	3–15
Sulfur (%)	0.5–2.5	0.5–6	0.3–1.5	0.3–2.5
Hydrogen (%)	1.5–3.5	4.5–6	5.5–6.5	6–7.5
Carbon (%)	75–85	65–80	55–70	35–45
Nitrogen (%)	0.5–1	0.5–2.5	0.8–1.5	0.6–1.0
Oxygen (%)	5.5–9	4.5–10	15–30	38–48
Btu/lb	12,000–13,500	12,000–14,500	7,500–10,000	6,000–7,500
Density (g/ml)	1.35–1.70	1.28–1.35	1.35–1.40	1.40–1.45

precision limits of ASTM standard methods. The discussion of the analysis of coal in this chapter pertains primarily to the ASTM standard methods (2).

5.1 SAMPLING

Preliminary to any laboratory testing of coal, it is imperative that a representative sample be obtained; otherwise, the most carefully conducted analysis is meaningless. Reliable sampling of a complex mixture such as coal is difficult, and handling and preparation of the sample for analysis presents further problems. Variations in coal handling facilities make it impossible to publish a set of rules that would apply to every sampling situation. The proper collection of the sample involves an understanding and consideration of the minimum number and weight of increments, the particle size distribution of the coal, the physical character and variability of the constituents of coal, and the desired precision.

Guidelines for the collection of gross samples of coal are given in ASTM method D 2234. A *gross sample* is defined as a sample representing a quantity, or *lot,* of coal and is composed of a number of increments on which neither reduction nor division has been performed. The recommended maximum quantity of coal to be represented by one gross sample is 10,000 tons (9,080 megagram). Ash content is the property most often used in evaluating sampling procedures. The ASTM general purpose sampling procedures are designed to give a precision such that if gross samples are taken repeatedly from a lot or consignment and one ash determination is made on the analysis sample from each gross sample, 95 out of 100 of these determinations will fall within ± 10% of the average of all determinations. When other precision limits are required or when other constituents are used to specify precision, some special-purpose sampling procedures are used.

5.1.1 Preparation of a Sample for Analysis

Once a gross sample has been taken, it is reduced in both particle size and quantity to yield a *laboratory sample.* The particle size distribution, or *size consist,* of the laboratory sample depends on its intended use in the laboratory and the nature of the tests to be run. The minimum allowable weight of the sample at any stage of reduction depends on the size consist, the variability of the constituents sought, and the degree of precision desired. Recommended minimum weights for group A coals (which have been cleaned in all sizes) and group B coals (all others, including unknown coals) are listed in Table 5.2.

The subsample reduced to 100% through a number 60 (250 μm) sieve and divided to not less than 50 g is called the *analysis sample,* which is required for most ASTM laboratory tests. The ASTM method D 2013 presents standard procedures for preparing coal samples for analysis. The steps followed preparing an analysis sample from a gross sample are given in Figure 5.1.

Many problems, such as the loss or gain of moisture, improper mixing of constituents, improper crushing and grinding, contamination of the sample by equipment, and oxidation of coal, may arise during the sampling and sample preparation processes. To minimize the moisture problem, all standard methods include, when necessary, an air-drying stage in the preparation of the analysis sample so that subsequent handling and analysis will be made on a laboratory sample relatively stable with reference to gain or loss of moisture from or to the laboratory atmosphere. In collecting, handling, reducing, and dividing the gross sample, all operations should be done rapidly and in as few steps as possible to guard against moisture loss or gain.

The distribution of mineral matter in coal presents problems for the crushing, grinding, and uniform mixing at each step of the sampling procedure. The various densities of the materials found in coal can easily cause their segrega-

TABLE 5.2 Preparation of a Laboratory Sample

Crush to Pass at Least 95% through Sieve	Divide to a Minimum Weight of g[a]	
	Group A	Group B
Number 4 (4.75 mm)	2,000	4,000
Number 8 (2.36 mm)	500	1,000
Number 20 (850 μm)	250	500
Number 60 (250 μm; 100% through)	50	50

Source: Reference 2, method D 2013; reprinted with permission from *Annual Book of ASTM Standards,* vol. 05.05, copyright ASTM, 1916 Race St., Philadelphia, PA 19103.

[a] If a moisture sample is required, increase the quantity of the number 4 (4.75 mm) or the number 8 (2.36 mm) sieve subsample by 500 g.

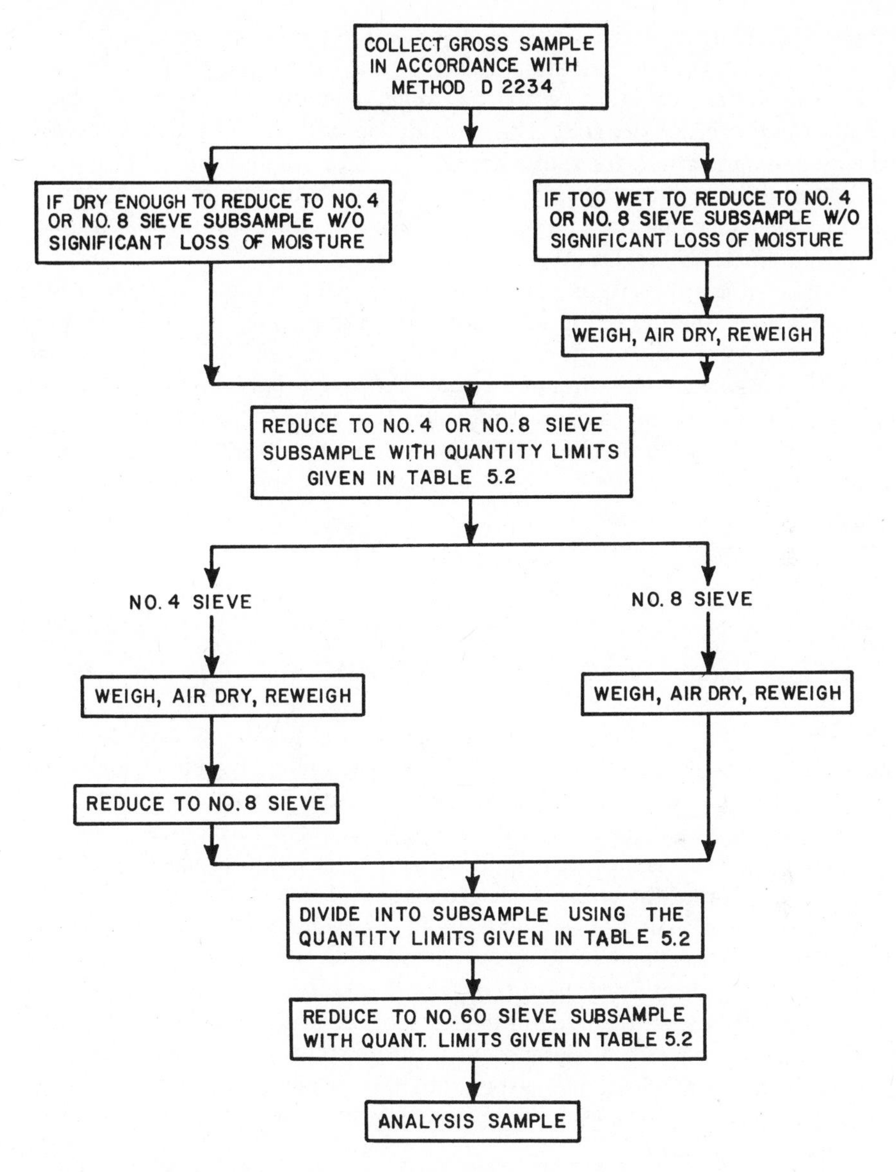

FIGURE 5.1 Analysis sample preparation: nonreferee method.

tion, especially if there is a wide range of particle sizes. Crushing and/or grinding coal from a large particle to a very small particle in one operation tends to produce a wide range of particle sizes and a high concentration of very fine particles. The crushing, grinding, and pulverizing should involve a reasonable number of steps, considering the starting particle size and nature of the coal. At the same time, it should be kept in mind that too many handling steps will increase the exposure of the coal to air and increase the chance of moisture changes and oxidation. Some models of coal sampling equipment give a wider range of particle sizes than others, due to the manner in which they crush and grind the coal. This should also be taken into consideration when planning routines for sample preparation. In addition to the problems already mentioned that may arise from the crushing and grinding operations, there is the chance that the equipment used may introduce some materials that will contaminate the coal sample.

Coal is susceptible to oxidation at room temperature. Like moisture changes, such oxidation has to be considered in sampling, preparing, and storing samples. All these operations should be done rapidly and in as few steps as possible to minimize oxidation of the coal. The sample containers used should have airtight lids to guard against moisture loss and exposure of the coal to air. Containers should be selected that will hold only the required amount of sample and leave a minimum of air space. Even when such precautions are taken, the samples change very quickly, so the analysis of a sample should be carried out as soon as possible after it is received.

5.2 COAL ANALYSIS

Coal analysis may be divided into three categories: proximate analysis, ultimate analysis, and miscellaneous analysis. Proximate analysis is the determination, by prescribed methods, of moisture, volatile matter, ash, and fixed carbon (by difference) content. Ultimate analysis involves the determination of carbon and hydrogen content as found in the gaseous products of the complete combustion of the coal; the determination of total sulfur, nitrogen, and ash content in the material as a whole; and the estimation of oxygen content by difference. Miscellaneous analysis is a collective category for various types of physical and chemical tests for coal that are commonly requested by coal producers and buyers. Included in this category are such tests as the determination of calorific value, analysis of the forms of sulfur present, chlorine analysis, trace metals analysis, and carbon dioxide analysis and determinations of free-swelling index (FSI), grindability, plastic properties by Gieseler plastometer, and ash fusibility.

Only an outline and a general discussion of each of the ASTM standard methods of coal analysis are given in this chapter. For precise details of the methods, it is necessary to refer to the latest edition of the *Annual Book of ASTM Standards*. In this chapter, the discussion of each method includes such topics as the nature of the constituents of the coal being analyzed; the chemical reactions that may take place during analysis; some of the difficulties encountered in the tests; and interpretation, uses, and limitations of the data obtained.

5.3 PROXIMATE ANALYSIS

The proximate analysis of coal was developed as a simple means of determining the distribution of products obtained when the coal sample is heated under specified conditions. It separates the products into four groups: (1) moisture; (2) volatile matter, consisting of gases and vapors driven off during pyrolysis; (3) fixed carbon, the nonvolatile fraction of coal; and (4) ash, the inorganic residue remaining after combustion. Proximate analysis is the most often used type of analysis for characterizing coals in connection with their utilization. Differences in the type of information desired and used by coal producers and consumers have led to variations in the kind and number of tests used under the rubric *proximate analysis*. Other terms used in the coal industry are *short prox* and *prox*. Common usage in the field tends to favor short prox, which is the determination of moisture, ash, Btu, and sulfur, while prox means the determination of moisture, ash, volatile matter, fixed carbon, Btu, and sulfur. Prox may also include ash fusion temperature and the free-swelling index of coal. Proximate analysis as defined by ASTM is the topic of this section.

5.3.1 Moisture

The most elusive constituent of coal to be measured in the laboratory is moisture. The moisture in coal ranges from 2–15% in bituminous coal to nearly 45% in lignite. There are several sources of the water that is found in coal. The vegetation from which coal was formed had a high percentage of water that was both physically and chemically bound. Varying amounts of water were still present at different stages of the coalification process. The overall result of coalification was to eliminate much of the water, particularly in the later stages, as is evident from a comparison of the moisture contents of different ranks of coal from lignite to anthracite (see Table 5.1). Water is present in most mines and circulates through most coal seams. After mining, many coals are washed with water during preparation for market and are then subject to rain and snow during transportation and storage. All of these sources contribute to the moisture in coal and to the problems associated with the measurement of this moisture.

The difficulties in accurately assessing the moisture in coal stem from the fact that the moisture exists in several forms. This water may be divided into four categories: inherent moisture, surface moisture, decomposition moisture, and water of hydration of mineral matter. *Inherent moisture* is also referred to as bed moisture or equilibrium moisture and is believed to be the water held in the capillaries of varying radii that are found in coal. The vapor pressure of this water is somewhat less than that of the moisture found on the surfaces of coal, which is appropriately called surface moisture or free moisture. *Surface moisture* has a vapor pressure equal to that of free water at the same temperature. *Decomposition moisture* is produced from the thermal decomposition of organic constituents of coal. The *water of hydration of mineral matter* is incorporated into the crystal lattices of the inorganic and claylike materials found in coal. Air drying removes most of the surface moisture of coal, while a temperature of approximately 107°C is needed to remove inherent moisture. At temperatures of approximately 200–225°C, moisture from the decomposition of organic materials is driven off, but water of hydration requires a considerable amount of energy for expulsion. For example, the water of hydration in kaolinite is not released until a temperature of about 500°C is reached (3). Decomposition moisture and water of hydration of mineral matter are not commonly dealt with in ordinary coal analysis because the temperatures used for routine moisture testing are well below those needed to remove these two kinds of moisture

In practice, the various forms of moisture in coal are described according to the manner in which they are measured by some prescribed standard method. These forms are inherent moisture, free or surface moisture, total moisture, air-dry loss moisture, residual or air-dried moisture, and as-received moisture. *Total moisture* is defined as ASTM standard D 121, as "that moisture determined as the loss in weight in an air atmosphere under rigidly controlled conditions of temperature, time and air flow as established in method D 3302." *Air-dry loss moisture* is the loss in weight resulting from the partial drying of coal, and *residual moisture* is that remaining in the sample after determining the air-dry loss moisture. Total moisture is the sum of the inherent and free moisture in coal and is also the sum of the air-dry loss and residual moisture. However, inherent moisture is not the same as residual moisture, nor is free moisture equivalent to air-dry loss moisture. Some relationships may be established between the quantities of inherent moisture and surface moisture, as they are determined by standard methods and the presence of these forms of water in coal. However, air-dry loss and residual moisture are determined as steps in an analytical procedure and should not be used as significant values for interpretation. It would simply be a coincidence if inherent moisture had the same value as residual moisture or if free moisture had the same value as air-dry loss moisture for a given coal sample. *As-received moisture* also is equal to the total moisture, or is the sum of the inherent and free moisture present in the coal at the time of the analysis.

Determination of Moisture. Many methods have been developed for determining the moisture content of coal. Most of these methods can be included in the following categories: (1) thermal drying methods, (2) desiccator methods, (3) distillation methods, (4) extraction and solution methods, (5) chemical methods, and (6) electrical methods (4). Most common tests for moisture involve a thermal drying procedure, usually at a temperature a few degrees above the boiling point of water; the moisture released upon heating is measured either directly or indirectly. The direct method involves the gain in weight of a weighing tube packed with desiccant through which the gases evolved from heating a coal sample are passed. This is probably the more accurate method, since only water is absorbed by the tube, although other gases, such as methane, are evolved. The indirect method is more often used, primarily because it is easier to do. The moisture is determined by measuring the weight loss of a coal sample upon heating in various atmospheres. If the coal is susceptible to oxidation, as most low-rank coals with a high moisture content are, the heating should be done in an inert atmosphere. The drying of most high-rank coals in air is accepted practice today.

Desiccator methods of determining moisture involve the determination of the loss in weight of a coal sample in the presence of a desiccant. Either a normal or reduced pressure (vacuum desiccator) may be used, but the drying is carried out at room temperature.

After thermal drying methods, distillation methods are the next most commonly used. In these procedures, coal is heated in a liquid that has a boiling point higher than that of water and is immiscible with it. Xylene, toluene, or a petroleum fraction of a selected boiling range are the liquids normally used. The distilled vapors are condensed in a graduated tube, and the volume of water is measured after the two liquids separate. This method is considered particularly advantageous for use with low-rank coals, since air is excluded from the coal, which minimizes the error due to oxidation. This is also a direct method of measuring moisture, and consequently there is no error due to the loss of other gases.

A nonthermal method of determining moisture involves the use of an extraction procedure in which the coal is shaken with a solvent that extracts the water from the coal. The degree of change in some physical property of the solvent, such as density, is then used as a measure of the water extracted.

Chemical methods used for determining moisture include the application of the Karl Fischer method of determining water content, and the reaction of quicklime with water in coal and the subsequent measurement of the heat generated by the reaction. Electrical methods of measuring coal moisture involve the determination of the capacitances or the resistances of quantities of coal. These methods have been used by industry, particularly for moving streams of coal.

ASTM Standard Methods of Analysis of Moisture. The ASTM standard methods of determining the amount of moisture in coal are the following:

Method D 1412: equilibrium moisture of coal at 96–97% relative humidity and 30°C

Method D 2961: total moisture of coal reduced to number 8 (2.38-mm) top sieve size (limited purpose method)

Method D 3173: moisture in the analysis sample of coal and coke

Method D 3302: total moisture in coal

In addition to these, ASTM method D 2013, "Method of Preparing Coal Samples for Analysis," gives directions for air drying samples. The air-drying methods used in method D 2013 are the same as those used in method D 3302.

Routine moisture determinations are carried out according to specifications in ASTM method D 3173 or method D 3302, depending on the state of preparation and/or condition of the coal sample. The entire procedure for determining the total moisture in coal, after collecting the gross sample, begins with preparing the sample for analysis, as outlined in method D 2013 (Fig. 5.1.) In routine work, if the gross sample is dry enough, it is reduced to number 4 or number 8 top size (number 4 top size means more than 95% of the sample passes through a number 4 sieve). If the sample is too wet to reduce in size, it is weighed before reduction. Air drying is carried out on a drying floor or in a special drying oven operated at 10–15°C above room temperature. The purpose of air drying is to reduce the moisture in the sample to approximate equilibrium with the air in the laboratory. This minimizes changes in moisture content when the sample is handled during the crushing and grinding operations or during an analysis. After reduction of the gross sample to number 4 or number 8 top size, it is divided and a laboratory sample taken. The laboratory sample is then air dried and reduced to number 8 top size if necessary. If the total moisture is to be determined as in method D 3302, number 8 top size coal is used, and residual moisture is determined by heating at 104–110°C for $1\frac{1}{2}$ h. If a full analysis of the coal (proximate or ultimate analysis) is desired, the laboratory sample must be reduced to number 60 (250-μm) size and divided, and an analysis sample must be taken. Using the analysis sample, residual moisture is determined according to method D 3173 by heating for 1 h at 104–110°C.

The moisture values obtained from the various drying procedures are expressed as a percentage, by weight, of the sample used in the particular test. Consequently, a correction factor must be used to make the various moisture values additive so that total moisture values can be obtained. The air-dry loss moisture and total moisture values can be calculated using the following formulas (with all values expressed in weight percent):

$$ADL = A' \times \frac{100 - A}{100} + A \qquad (1)$$

$$M = R \times \frac{100 - ADL}{100} + ADL \qquad (2)$$

where ADL = complete air-dry loss moisture, A' = air-dry loss of laboratory sample, A = air-dry loss of gross sample, M = total moisture of gross sample, and R = residual moisture.

The equilibrium moisture in coal as determined by method D 1412 probably gives the best value of the inherent moisture. In this method, a sample is brought into equilibrium in a partially evacuated desiccator with an atmosphere of 96–97% relative humidity at 30° C. The amount of moisture in the coal under these conditions is determined by weight loss upon heating.

As in all methods of determining moisture, there are problems associated with this equilibrium moisture method, and precautions must be taken to obtain reliable results. Overdried and/or oxidized coal result in low moisture values. To prevent overdrying, the sample should be kept wet before this test is run. Nothing can be done for samples that are oxidized before testing, but oxidation of the coal during the test can be prevented by using a dry nitrogen atmosphere. This procedure is particularly useful for low-rank coals. During the test itself, it is important to observe the specified temperature and time limits for equilibration and restoring the pressure in the desiccator to atmospheric conditions. A sudden lowering of the temperature or a sudden surge of air into the desiccator after equilibration may cause condensation of moisture on the coal. Mechanical losses of the coal sample caused by sudden surges of air into the evacuated desiccator when atmospheric pressure is restored of course void the results of the test.

In routine moisture determinations, sample handling should be kept at a minimum, since loss or gain of moisture may occur during prolonged handling. If too long a period is used in completing the analysis of a coal sample, moisture may evaporate from the coal in a container and condense on the container. It is almost impossible to redistribute this moisture uniformly once this has occurred. Changes in the moisture content may also occur during reduction of the gross sample. Heat generated by the crushing and grinding operations may be sufficient to cause moisture loss. The relative humidity of the sample preparation and laboratory rooms is likely to be different from the atmosphere where the gross sampling was done. The relative humidity in the laboratory rooms also may change while a complete analysis is being carried out. Air-drying steps in the analysis and efficient sample handling help minimize the effects of relative humidity changes.

Exposure of the coal sample to the atmosphere for extended periods of time increase the chances of oxidation, which would result in a weight gain by the coal sample that would offset part of the loss of moisture and give moisture results that are low and misleading. In the determination of moisture by a weight loss method, it is necessary to attain a constant weight, which requires alternate heating and cooling of samples. Prolonged heating or an excessive number of alternate heating and coolings should be avoided to minimize the chances of oxidation.

Interpretation and Uses of Moisture Data. Moisture values are very important due to the influence they have on other measured and calculated values used in coal analysis and, ultimately, to the part they play in the buying and selling of coal. The various forms of moisture in coal and the methods by which moisture values are obtained have been discussed in the preceding sections. The interpretation of moisture data and the uses and limitations of these data is of primary concern to the analyst.

Usually the first moisture value to be obtained on a coal sample is the air-dry loss moisture. This moisture loss occurs during an attempt to bring the coal sample into equilibrium with the atmosphere in the sample preparation room. Temperatures used for air drying vary over a wide range. The ASTM specifications call for air drying on a drying floor at room temperature or in a drying oven at temperatures 10–15°C above room temperature, with a maximum of 40°C. The practice of using temperatures above room temperature may accelerate oxidation but shortens the time needed for air drying, which reduces total exposure of the coal and decreases the chances of oxidation. The shorter exposure time should compensate for the use of the elevated temperature. In very warm climates or on very warm days in moderate climates, it may not be possible to conduct air-drying experiments without exceeding the recommended maximum temperature. Temperatures much above 40–50°C, however, should not be used for air drying.

The air-dry loss moisture as a percentage of the total moisture in coal is variable. It may vary from 25–90% of the total moisture for different samples and may vary widely for coals of the same rank. It has been used incorrectly in some instances as a measure of the surface or free moisture. The use of the air-dry loss moisture value by itself has no real significance in the characterization of coals.

Residual moisture, or as-determined moisture, is used to calculate other measured analytical values to the dry basis. Residual moisture alone has no significance in the characterization of coals.

The sum of residual moisture and air-dry loss moisture is equal to the total moisture. As measured in ASTM method D 3302, total moisture in coal is that which exists at the site, at the time, and under the conditions it is sampled. It applies to coals as mined, processed, shipped, or utilized in normal commercial operations. Coal-water slurries, sludges, or pulverized products under 0.5 mm diameter sieve size are exceptions. Total moisture applies to coals of all ranks, with consideration of oxidation and decomposition variations that may occur in lower-rank coals.

The total moisture is used for calculating other measured quantities to the as-received basis. In the buying and selling of coal, as-received calorific values are often used as the basis for contracts. To obtain as-received calorific values, the experimentally determined dry calorific values are converted to the as-received basis using total moisture values. When thousands of tons of coal are involved in a contract, an error that may seem insignificant in a normal laboratory situation may be serious from a monetary standpoint.

Total moisture is important in assessing and controlling the commercial processing of coals. It is used to determine the amount of drying that is needed to reach a given moisture requirement and to determine the amount of dust-proofing and freeze-proofing agents to add. In coking processes, coals with a high moisture content require more heat for vaporization of the moisture, which leads to longer coking cycles and decreased production. The total moisture of the coal used must be accurately known to allow for proper charging of the coke ovens and overall control of the coking process.

Inherent or equilibrium moisture is used for calculating moist mineral-matter-free calorific values for the rank classification of high-volatile bituminous coals. It is also used for estimating free or surface moisture, since total moisture is equal to the sum of the inherent moisture and the free moisture. This moisture value is also referred to as bed moisture, since it is considered the inherent moisture of the coal as it occurs in the unexposed seam, where the relative humidity is probably near 100%. The primary reason for using a high relative humidity in the determination of equilibrium moisture is to approximate 100% relative humidity. However, due to physical limitations, equilibrium moisture determinations are made at 96–97% relative humidity and used as inherent moisture values. It has been found that equilibrium moisture determined at 96.7% relative humidity and 30°C averages about 95% of the extrapolated 100% value, and equilibrium moisture determined at 97.7% relative humidity and 30°C averages about 96% of the 100% value. These values were based on data for the three ranks high-volatile A, B, and C bituminous coal found in Illinois (5).

The banded constituents—vitrain, clarain, durain, and fusain—that occur in coal vary considerably in the amount of moisture they hold at various relative humidities. One constituent, fusain, holds relatively little moisture below 90% relative humidity. This is another reason for using a high relative humidity in the determination of the equilibrium moisture of coal.

Surface moisture values are really estimates. These are obtained by subtracting equilibrium moisture from total moisture. However, there is no sharp dividing line between inherent moisture and surface moisture. The measurement of inherent moisture depends on the fact that its vapor pressure is less than that of surface moisture. It is commonly thought that inherent moisture is contained in the pores and capillaries of coal. However, these pores and capillaries may vary in diameter and size to such an extent that the water in the larger capillaries has a vapor pressure approaching that of surface moisture.

Drying, pulverizing, dust-proofing, and the general handling of coal all depend on surface moisture data. Too much surface moisture is particularly troublesome in pulverizing and handling operations. A wet coal is very difficult, and in some instances almost impossible, to pulverize. The presence of only 0.5% surface moisture is enough to cause coal to stick in a chute.

There is no simple and reliable method of determining the water of hydration of mineral matter. The average value of 8% of the ash is used as the value for water of hydration of mineral matter in coals in the United States. This value is acceptable, although it is an average of values that range from 2–3% up to 15–20%. Water of hydration values are used to correct ash to the form of hydrated minerals in mineral matter calculations.

5.3.2 Ash

Coal ash is the residue remaining after the combustion of coal under specified conditions. It does not occur as such in the coal but is formed as the result of chemical changes that take place in the mineral matter during the ashing process. The quantity of ash can be more than, equal to, or less than the quantity of mineral matter in coal, depending on the nature of the mineral matter and the chemical changes that take place in ashing.

There are two types of ash-forming materials in coal: extraneous mineral matter and inherent ash. The *extraneous mineral matter* consists of materials such as calcium, magnesium, and ferrous carbonates; pyrite; marcasite; clays; shales; sand; and gypsum. *Inherent mineral matter* represents the inorganic elements combined with organic components of coal. The origin of such materials is probably the plant materials from which the coal was formed. Ash from the inherent mineral matter is usually insignificant as far as the total quantity of ash is concerned (6).

The composition of coal ash varies widely, depending on the mineral matter associated with the coal. Typical limits of the composition of ash of bituminous coals are given in Table 5.3.

Some of the chemical changes that take place during ashing include the loss of water of hydration from claylike material, the loss of carbon dioxide from mineral carbonates, and the conversion of pyrite to Fe_2O_3 and oxides of sulfur. The organically combined inorganic elements are probably converted to oxides as well. Some recombination reactions do occur, depending on the conditions and composition of materials in the coal sample. Under certain conditions, sulfur dioxide is oxidized to SO_3, which reacts with metal oxides, particularly calcium oxide, to form stable sulfates, which remain in the ash.

Determination of Ash Content. The procedure for the determination of ash content in coal is outlined in ASTM method D 3174. In this method, 1 g of coal is weighed and placed in the preweighed porcelain crucible. The coal used may be from the analysis sample or the dried coal sample from the moisture determination, method D 3173. The crucible is then placed in a cold muffle

TABLE 5.3 Typical Limits of Ash Composition of Bituminous Coals

Constituent	United States	England	Germany
SiO_2	20–60	25–50	25–45
Al_2O_3	10–35	20–40	15–21
Fe_2O_3	5–35	0–30	20–45
CaO	1–20	1–10	2–4
MgO	0.3–4	0.5–5	0.5–1
TiO_2	0.5–2.5	0–3	—
$Na_2O + K_2O$	1–4	1–6	—
SO_3	0.1–12	1–12	4–10

Source: Reference 7; reprinted with permission from H.H. Lowry, ed., *Chemistry of Coal Utilization,* supplementary vol., John Wiley & Sons, Inc., 1963, p. 209.

furnace, and the temperature is raised at such a rate that it reaches 450–500°C in 1 h. Heating is continued so that a temperature of 700–750°C is reached after 2 h, and the furnace is held at this temperature for an additional 2 h. During the ashing procedure, an ample supply of air (two to four volume changes per minute) must be supplied to the furnace.

Coals with unusually high amounts of calcite ($CaCO_3$) and pyrite may retain varying amounts of sulfate sulfur upon ashing. Coals are ashed by the two-step heating rate in an attempt to minimize this retention. Pyrites are oxidized to sulfur oxides and iron oxides at temperatures around 450–500°C. Calcite and other carbonate minerals decompose to the metal oxides and carbon dioxide at temperatures near 700°C. Oxidizing pyrites at the lower temperatures rids the sample of sulfur that may be converted to sulfur oxides and retained by the metal oxides formed at the higher temperatures.

Interpretation and Uses of Ash Data. The value obtained for the ash content is not a true indication of the noncombustible material occurring in coal. The indefinite amount of sulfur that may be retained in the ash and the high temperature interaction of the various ash-forming components to produce new compounds make it impossible to give an exact interpretation of the relationship between the composition of the ash and clinkering or fusing of ash particles, boiler tube slagging, and other problems associated with ash formation. The ash value is an empirical quantity, but it is quite useful for many practical applications.

The ash value is the analytical value most commonly used for evaluating sampling procedures and is one of the values normally specified in coal contracts. In combustion, high ash content reduces the amount of heat obtainable from a given quantity of coal. High ash content also leads to the problem of handling and disposing of larger amounts of ash produced during combustion. The composition of coal ash is considered in the amount of clinkering and boiler tube slagging that may occur in a boiler. The design of most boilers is such that only coals with a specified range of ash content may be used in the efficient operation of the boiler. The amount of ash in coal used in a coking process is an indication of the amount of ash that will remain in the coke that is made. Coke with a high ash content that is used in a blast furnace requires more fluxing limestone to compensate for the ash, and a greater volume of coke to obtain the required amount of usable carbon.

Coal can be cleaned by various processes to reduce the ash and sulfur content. The ash content of raw coal is often used to select the best cleaning method, and the ash content of the cleaned coal is used to measure the effectiveness of the cleaning process. In the commercial pulverization of coals, the amount and nature of ash is carefully considered before selecting pulverizing equipment or setting up the process.

Finally, in the American system of classifying coals by rank, it is necessary that some of the parameters that are used be calculated to a mineral-matter–free basis. The ash value is needed for these calculations.

5.3.3 Volatile Matter

The loss of mass, corrected for moisture, that results when coal is heated in specified equipment under prescribed conditions is referred to as volatile matter. The matter lost is composed of materials that form upon the thermal decomposition of the various components of coal. Some of the constituents of coal volatile matter are hydrogen, carbon monoxide, methane and other hydrocarbons, tar vapors, ammonia, some organic-sulfur- and oxygen-containing compounds, and some incombustible gases, such as carbon dioxide and water vapor, all of which come from the decomposition of organic materials in coal. Inorganic materials in coal contribute the water of hydration of mineral matter, carbon dioxide from carbonates, and hydrogen chloride from inorganic chlorides to the volatile matter. Volatile matter does not include the residual moisture, as determined in method D 3173.

Determination of Volatile Matter Content. Volatile matter is determined by establishing the loss in weight resulting from heating a coal sample under rigidly controlled conditions. The ASTM procedure for determining volatile matter is outlined in method D 3175. The method is an empirical one that requires close adherence to detailed specifications. The type of heating equipment (electric Fieldner-type tube furnace) and the size and shape of the sample holders as well as the material from which they are made (platinum crucibles specified) all have some influence on the rate of heating of the sample and the range of temperatures to which it is exposed. The crucibles used are 10–20 ml capacity of specified size with deep-fitting lids.

There are two procedures. The regular method is used for nonsparking coals and coke. The modified method is used for fuels that do not yield a coherent cake as residue in the determination and that evolve gaseous products at a rate sufficient to carry solid particles out of the crucible when heated at the standard rate. Such fuels are referred to as sparking fuels and normally include all low-rank noncaking coals and lignites, but may include other coals as well.

In the regular procedure, 1 g of the analysis sample of coal is weighed and placed in a preweighed platinum crucible (10–20 ml capacity, 25–35 mm in diameter, and 30–35 mm in height) with a close-fitting cover. The crucible is then suspended at a specified height in the furnace chamber. The temperature of the region in the furnace where the crucible is suspended must be maintained at 950°C ± 20°C. After the more rapid discharge of volatile matter, as evidenced by the disappearance of the luminous flame, the cover of the crucible should be tapped to ensure that the lid is still properly seated to guard against the admission of air. After heating for exactly 7 min, the crucible is removed from the furnace and cooled. The crucible should be weighed as soon as it is cold. The percentage loss of weight minus the percentage moisture equals the volatile matter.

In the modified procedure for all sparking fuels, the sample is suspended and heated in a cooler zone of the furnace such that the temperature inside the crucible reaches 600° C ± 50° C in 6 min. After the preliminary heating, the crucible is lowered into the hot zone (950° C ± 20° C) of the furnace and held there for 6 min. The crucible is then removed from the furnace and set on a metal block to cool before weighing. The cooling period should be kept constant and should not exceed 15 min to ensure uniformity of results. The volatile matter is calculated in the same manner as in the regular method.

The furnace used for the volatile matter determination must be checked frequently because any variance from proper standardization produces erratic results. The thermocouple in the furnace chamber may break or change over a period of time. Probably no two furnaces have the same heating characteristics. It is necessary, therefore, to check them occasionally, especially after any repairs.

The rate of heating of the sample influences volatile matter values and makes it necessary to calibrate equipment to achieve a satisfactory and reproducible heating rate. This calibration can be accomplished by using either a manual or an automatic mechanical device that lowers the sample crucible at a reproducible rate into the electrically heated furnace.

Sparking is caused by incandescent particles of coal that are carried out of the crucible by the rapid release of moisture or volatile matter. The loss of these particles results in volatile matter values that are too high. Sparking may increase with an increase in the amount of very fine particles in the analysis sample. The concentration of such fine particles can be avoided to some degree by proper reduction of coal particles during the preparation of the analysis sample.

The crucibles and covers must be properly shaped to ensure a proper fit. A loose-fitting cover allows air to come in contact with the hot coal sample, with subsequent formation of volatile carbon oxides. Such an occurrence would result in a volatile matter value that is too high. Oxidation is not a serious problem in volatile matter determinations, since the rapid release of large amounts of gases during the test does prevent the entry of air into the crucible, thereby reducing the chance of oxidation. Addition of a few drops of a volatile material, such as toluene, may also help prevent oxidation.

The mass lost in the ASTM procedure for determining the volatile matter in coal includes the residual moisture and the water of hydration of mineral matter. The water of hydration of mineral matter is therefore included in the volatile matter, since there is no satisfactory method of determining it. However, the residual moisture value is subtracted from the mass lost to obtain the determined volatile matter. The volatile matter values obtained by the ASTM procedure using dried coal samples are lower than those obtained using the analysis sample. Studies of bituminous coals in which the volatile matter values were determined using dried samples from the residual moisture determination (method D 3173) yielded values that were an average of 0.94% (absolute) lower than the corresponding ASTM values (8). The coals studied had residual moisture values ranging from 1.71% to over 10%. There was a moderate correlation between the residual moisture content of the coals and the difference between the volatile matter values.

Interpretation and Uses of Volatile Matter Values. In the determination of volatile matter content, the modified ASTM method, using a slower heating rate, is applicable to a wider variety of coals. However, the values obtained are sometimes lower (1–3% absolute) than those obtained from the regular method. This illustrates the empirical nature of this test and the importance of strict adherence to detailed specifications. The complexity of the constituents of coal that undergo decomposition during this test makes it necessary to have wide tolerances for reproducibility and repeatability.

Volatile matter values are important in choosing the best match between a specific type of coal burning equipment and the coal to use with the equipment. Such values are valuable to fuel engineers in setting up and maintaining proper burning rates. Volatile matter values are used as an indication of the amount of smoke that may be emitted from furnaces or other types of coal burning equipment. Limits may be set on the volatile matter content of the coal used in certain coal burning facilities in order to control smoke emissions.

Volatile matter values are also important in the selection of coals and in determining the blending proportions of coal for coking. The volatile matter value of coke is used as a means of evaluating the extent of coking, depending on the intended use of the coke.

5.3.4 Fixed Carbon

The fixed carbon value is obtained by subtracting the sum of the percentages of moisture, ash, and volatile matter from 100. This value is considered to be the amount of carbon residue that remains after the volatile matter test. The residue is the product of the thermal decomposition of the coal.

Interpretation and Uses of Fixed Carbon Data. The fixed carbon value is one of the values used in determining the efficiency of coal burning equipment. It is a measure of the solid combustible material that remains after the volatile matter in coal has been removed. For this reason, it is also used as an indication of the yield of coke in a coking process. Fixed carbon plus ash essentially represents the yield of coke. Fixed carbon values, corrected to a dry, mineral-matter–free basis, are used as parameters in the ASTM coal classification system.

5.4 ULTIMATE ANALYSIS

Ultimate analysis of coal and coke is defined as the determination of the carbon and hydrogen in the material, as found in the gaseous products of its complete combustion; the determination of sulfur, nitrogen, and ash in the material as a whole; and the estimation of oxygen by difference. The carbon determination includes that present in the organic coal substance and any originally present as mineral carbonate. The hydrogen determination includes that present in the organic materials in coal and in all water associated with the coal. All nitrogen determined is assumed to be part of the organic materials in coal. Sulfur occurs in three forms in coal: as organic sulfur compounds; as inorganic sulfides, which are mostly the iron sulfides pyrite and marcasite; and as inorganic sulfates. The total sulfur value is used for ultimate analysis.

Moisture is not by definition a part of the ultimate analysis of coal but must be determined so that the analytical values obtained can be converted to bases other than that of the analysis sample. In other words, analytical values may need to be converted to an as-received or a dry basis. When suitable corrections are made for the carbon, hydrogen, and sulfur derived from the inorganic material, and for conversion of ash to mineral matter, the ultimate analysis represents the elemental composition of the organic material in coal in terms of carbon, hydrogen, nitrogen, sulfur, and oxygen.

5.4.1 Carbon and Hydrogen

Almost all the carbon and hydrogen in coal occurs in the combined form. Both of these elements are present in the very complex organic compounds found in coal. Carbon also occurs in the mineral carbonates, with calcite being the principal component. Hydrogen is also present in the various forms of moisture found in coal.

Determination of Carbon and Hydrogen Content. All methods of determining the carbon and hydrogen content of coal are very similar in that a weighed sample is burned in a closed system under carefully controlled conditions. The carbon is converted to carbon dioxide and the hydrogen to water. In the ASTM method, the products of the combustion are absorbed in an absorption train after removal of any interfering substances. The amount of carbon and hydrogen is calculated from the weight gained by the reagents in the absorption train.

The equipment used for the determination of the carbon and hydrogen content of coals basically consists of two parts: the combustion unit and the absorption train. According to ASTM method D 3178, the combustion unit is made up of three separate electrically heated furnaces and the combustion tube. The first furnace, at the inlet end of the tube, is approximately 130 mm long and can be moved along the tube. The second and third furnaces are 330 mm and 230 mm in length, respectively, and are mounted in a fixed position around the combustion tube. The required operating temperatures are 850–900°C, 850°C ± 20°C, and 500°C ± 20°C for the first, second, and third furnaces, respectively. The combustion tube can be constructed from fused quartz or high-silica glass. The dimensions of the tube are 19–22 mm inside diameter and a minimum length of 970 mm. The exit end is tapered to allow connection to the absorption train. The combustion tube is packed with lead chromate or silver gauze under the third furnace and cupric oxide under the second furnace. Oxidized copper gauze plugs are used to contain these components in the tube. A diagram of the arrangement of tube fillings appears in Figure 5.2.

The absorption train is composed of a tube packed with a water absorbent, a second tube packed with a carbon dioxide absorbent, and a guard tube packed with equal volumes of the water and carbon dioxide absorbents. Anhydrous magnesium perchlorate is commonly used for the water absorbent, while sodium or potassium hydroxide impregnated in an inert carrier is used as the carbon dioxide absorber.

In the analysis of a coal sample, 0.2 g of the analysis sample is placed in either a glazed porcelain, fused silica, or platinum boat and inserted into the combustion tube under the first furnace. The sample is burned in an oxygen atmosphere, and the combustion products are allowed to flow over the heated copper oxide and lead chromate or silver and into the absorption train. The copper oxide catalyzes the complete combustion of the carbon and hydrogen in the coal, whereas the lead chromate absorbs the oxides of sulfur. If silver gauze is used, both the sulfur oxides and chlorine will be absorbed. Water and carbon dioxide are absorbed by the preweighed absorbers in the absorption train, and the percentages of carbon and hydrogen in the sample are calculated from the gain in weight of absorbers.

Some problems that may arise with the use of the equipment described are the incomplete combustion or conversion of carbon to carbon dioxide and of hydrogen to water. Several things may go wrong and cause these problems to occur. If the unit used for burning the sample is heated too rapidly, volatile matter may be released at such a rate that some of it may pass through the entire system and not be completely burned or absorbed. To prevent this from happening, the temperature of the combustion unit must be at the proper level, and enough time must be allowed for complete combustion. In addition,

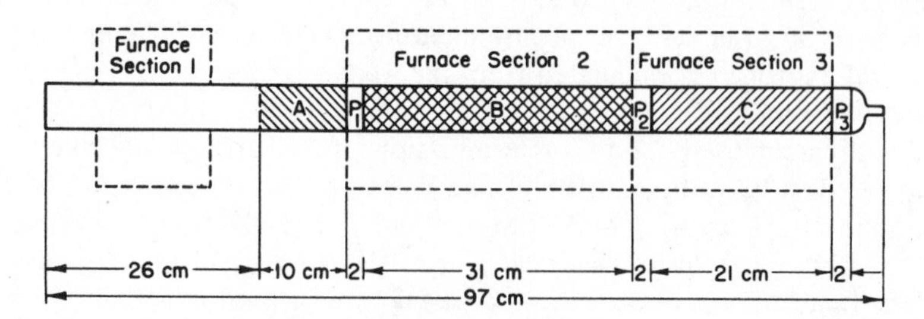

FIGURE 5.2 Arrangement of tube fillings for combustion tube. A: clear fused quartz section (optional) when a translucent quartz tube is used; B: cupric oxide filling; C: lead chromate or silver filling; P_1, P_2, or P_3: oxidized copper gauze plugs. (*Source:* Reprinted with permission from *Annual Book of ASTM Standards,* vol. 05.05, copyright ASTM, 1916 Race St., Philadelphia, PA 19103.)

a proper flow of oxygen must be maintained through the system, and all connections in the apparatus must be made gas tight.

The temperature of the third furnace, which covers the lead chromate and/or silver, must be maintained carefully. If it is too high, the lead chromate may fuse with the glass tubing or, if silver is used, some of the silver sulfate or chloride may be decomposed. The chlorine and/or oxides of sulfur that are released would then be absorbed in the absorption train, yielding high values for carbon and hydrogen.

Whenever a combustion tube is put into use after standing idle for some time, it is necessary that the tube be reconditioned before making any determinations. To condition the system, the combustion train is tested under normal operating conditions (heating for approximately 40 min), including the flow of oxygen. This testing is repeated until the absorption bulbs on successive weighings do not vary by more than 0.5 mg. Newly packed combustion tubes are conditioned by burning a coal sample under normal operating conditions. It is not necessary for the absorption bulbs to be in place for this.

In this method, all organic carbon is burned to carbon dioxide. Inorganic carbonates are also decomposed under the conditions used, and the CO_2 produced is absorbed in the absorption train. For coals that have a high carbonate content, it may be necessary to determine the carbonate carbon content and subtract it from the total carbon content so that a more accurate value of the combustible carbon content is obtained.

The formation of oxides of nitrogen during the combustion process may lead to slightly high results for carbon and hydrogen, since the oxides are acidic in nature and would be absorbed in the absorption train. An extensive study was conducted by five laboratories to determine whether errors due to oxides of nitrogen are large enough to make their removal necessary (9). The data from these laboratories indicated that the hydrogen value would not be greatly influenced but that the carbon value would be higher by as much as 0.35%. This amount of error would not be significant in ultimate analysis used for commercial purposes. For more precise results, such as in certain research applications, these oxides of nitrogen can be removed by absorption on manganese dioxide, or in some cases lead dioxide, before absorption of the water and carbon dioxide.

Interpretation and Uses of Carbon and Hydrogen Data. As mentioned in the previous section, inorganic carbonates and nitrogen oxides contribute to the carbon value in coal as it is normally determined. Hydrogen values also are usually high, due to the inclusion of the various forms of moisture that are present in coal. All of these factors limit the reliability of carbon and hydrogen data for predicting the amount of combustible carbon and hydrogen in coal.

A reasonable correction to the hydrogen value for the moisture in coal can be made by subtracting one-ninth of the determined moisture from the determined hydrogen. A correction to the hydrogen value for the water of hydration of mineral matter is more difficult. The water of hydration of mineral matter for coals in the United States has been estimated to be 8% of the ash value. A correction to the hydrogen value for the water of hydration can thus be estimated by multiplying the ash value by 0.08, and one-ninth of this figure will give the correction to be subtracted from the determined hydrogen. Upon making these corrections for the forms of moisture, the value for the hydrogen in the analysis sample of coal is given by

$$\mathrm{H} = \mathrm{H}\text{ (as-determined)} - 2.02/18.02\,[M\text{ (as-determined)} + 8/100\,A\text{ (as-determined)}] \tag{3}$$

The results of the carbon and hydrogen analysis may be reported on any number of bases, differing from each other in the manner by which moisture values are treated. Inclusion of the hydrogen of moisture and water of hydration of mineral matter in the hydrogen value is common practice for the as-determined and as-received bases. Hydrogen values on the dry coal basis, however, are commonly corrected for the hydrogen of moisture. No corrections are normally made to the determined hydrogen value for the water of hydration of mineral matter, due to the uncertainty of the estimate of its value. Examples of the calculations involved and equations used in the treatment of ordinary laboratory data are given in Section 5.5.

Carbon and hydrogen values are used in the calculation of heat values for checking determined calorific values and for calculating heat balances in boiler efficiency tests. Probably the most common use of carbon and hydrogen data is for basic coal research.

5.4.2 Nitrogen

Nitrogen occurs almost exclusively in the organic matter of coal. Very little information is available concerning the nitrogen-containing compounds present in coal, but they do appear to be stable and are thought to be primarily heterocyclic. The original source of nitrogen in coal may have been both plant and animal protein. Plant alkaloids, chlorophyll, and other porphyrins contain nitrogen in cyclic structures stable enough to have withstood changes during the coalification process and thus to have contributed to the nitrogen content of coal.

Determination of Nitrogen. The best-known methods of determining nitrogen in solid fuels involve its liberation in measurable form from the organic material in which it occurs. The methods are based on quantitative chemical reactions and involve converting the nitrogen into ammonia or oxidizing it to the elemental state.

The Kjeldahl-Gunning macromethod is the one most widely used for determining nitrogen and is recognized as the current ASTM standard method (method D 3179). By this method, any nitrogen present in the sample is converted into ammonium salts by the destructive digestion of the sample by a hot mixture of concentrated sulfuric acid and potassium sulfate. After the digestion mixture has been made alkaline with sodium or potassium hydroxide, ammonia is expelled by distillation, condensed, and absorbed in an excess of acid. The determination of ammonia is completed by titration of the excess acid. The ASTM alternative method is the same except that, after the complete digestion, the ammonia is distilled into a boric acid solution and titrated with a standard acid solution. Proper precautions should be taken in carrying out this procedure, especially the digesting and distillation steps. In addition to the possibility of losing nitrogen-containing species if the proper heating rate is not observed, there is the problem of working with hot concentrated sulfuric acid and caustic solutions.

A catalyst is used in the Kjeldahl-Gunning method to increase the rate of digestion of the nitrogen-containing sample and shorten the digestion period. The total digestion time for most bituminous and low-rank coal samples is 3–6 h, even with the aid of a catalyst, whereas anthracite and coke samples may require as much as 12–16 h. The catalyst used in this method may be one of the following: elemental mercury, mercuric sulfate ($HgSO_4$) and selenium, mercuric selenite ($HgSeO_3$), or cupric selenite dihydrate ($CuSeO_3 \cdot 2H_2O$). Whenever mercury or a mercury-containing catalyst is used, the addition of potassium or sodium sulfide to the digestion mixture is necessary. The sulfide ions precipitate any mercuric ions as mercuric sulfide and prevent them from forming a stable complex ion with the ammonia produced in the digestion.

The Kjeldahl-Gunning semimicromethod has been used for nitrogen analysis in some coal laboratories. Proponents of this method claim that the determination of nitrogen is speeded up significantly with no loss of accuracy (10). The primary differences between the semimicromethod and the macromethod are that in the semimicromethod smaller-sized equipment is used, smaller samples are analyzed, and after the digestion mixture is made alkaline, ammonia is separated by steam distillation. In the semimicromethod, 0.1-g samples are used, whereas 1-g samples are used in the macromethod.

The most serious problem associated with the use of the Kjeldahl-Gunning method is the incomplete conversion of nitrogen in the nitrogenous compounds to ammonia. Complete conversion may not be accomplished for several reasons. In the decomposition of the nitrogenous compounds, the nitrogen is converted or reduced to ammonia, and organic materials are oxidized to various products. The digestion rate can be increased by the addition of stronger oxidizing agents, which more readily oxidize the organic matter. However, this cannot be done under normal conditions because the nitrogen would also be oxidized to nitrogen oxides and be lost from the analysis. The reaction mixture is not capable of reducing nitrogen oxides or nitro compounds. The decomposition must be carried out within a very narrow oxidation-reduction range. In addition, pyridine carboxylic acids may be formed that are resistant to decomposition. Potassium sulfate is added to the sulfuric acid digestion mixture to raise its boiling point. At no time in the digestion process should the composition of the digestion mixture approach that of potassium acid sulfate, or ammonia will be lost. Because of the somewhat limited oxidation and digestion conditions, and the possible formation of unwanted but stable by-products, a lengthy digestion period is required.

In the analysis of a coal sample, the heterogeneous digestion mixture may become a clear, straw-colored solution. To ensure complete conversion of the nitrogen to ammonia, the digestion must be continued for an additional $1\frac{1}{2}$–2 h beyond the straw-colored stage. Finer grinding of the more resistant coals may shorten the digestion time (12–16 h for anthracite and coke). The addition of chromium(III) oxide (Cr_2O_3) to the digestion mixture increases the rate of digestion of coke.

Interpretation and Uses of Nitrogen Data. Nitrogen data are primarily used in research and for the comparison of coals. These values are needed so that the oxygen content of a coal can be estimated by difference. During combustion, the nitrogen in coal can be converted to ammonia, elemental nitrogen, or nitrogen oxides, depending on the conditions of burning and the nature of the coal used. Nitrogen values could possibly be used to estimate the amount of nitrogen oxides that would be emitted upon burning of certain coals. Coal nitrogen values are also useful in predicting the amount of nitrogen in the products of coal liquefaction and gasification processes.

5.4.3 Total Sulfur

Sulfur occurs in three forms in coal: as part of the organic matter; as inorganic sulfides, primarily pyrite and marcasite; and as inorganic sulfates. Free sulfur as such does not occur in coal to any significant extent (11). The amount of the sulfur-containing materials in coal varies considerably, especially for coals from different seams. This variation is not as great for coals from a given field. On the average, coals from the Illinois Basin contain about equal amounts of organic and inorganic sulfur, although the relative amounts of these two sulfur forms may make up as much as 20–80% of the total sulfur in individual coals. Most of the sulfur-containing organic compounds in coal are heterocyclic in nature and are likely to be uniformly distributed throughout the coal matrix. Pyrite and marcasite are two different crystal forms of FeS_2. For this reason, they are usually referred to simply as pyrite. Pyrite is not uniformly distributed in coal. It can occur as layers or slabs or may be disseminated throughout the organic material as very fine crystals of $\frac{1}{2}$–40 μm size. The content of sulfates, mainly gypsum ($CaSO_4 \cdot 7H_2O$) and ferrous sulfate ($FeSO_4 \cdot 7H_2O$), rarely exceeds a few hundredths of a percent, except in highly weathered or oxidized coals.

Determination of Total Sulfur. The sulfur content is an important value to consider in the utilization of coal and coke for most purposes. Much work has been done in improving the accuracy and precision of sulfur determinations and in reducing the time for the analysis.

There are two ASTM methods of determining the total sulfur in coal and coke, with alternative procedures in each method. ASTM method D 3177, "Total Sulfur in the Analysis Sample of Coal and Coke," has two alternative procedures referred to as the Eschka and the bomb washing methods. ASTM method D 4239, "Sulfur in the Analysis Sample of Coal and Coke Using High-Temperature Tube Furnace Combustion Methods," has three alternative procedures. The basic difference in the three procedures is the method of detection of the sulfur dioxide produced in the combustion of the coal or coke sample. Method A uses an acid-base titration for detection, method B uses as iodimetric titration, and method C uses the absorption of infrared radiation for detection.

In the Eschka method, 1 g of the analysis sample is thoroughly mixed with 3 g of Eschka mixture, which is a combination of two parts by weight of light calcined magnesium oxide with one part of anhydrous sodium carbonate. The combination of sample and Eschka mixture is placed in a porcelain crucible (30 ml) and covered with another gram of Eschka mixture. The crucible is placed in a muffle furnace, heated to a temperature of 800°C ± 25°C, and held at this temperature until oxidation of the sample is complete. The sulfur compounds evolved during combustion react with the MgO and Na_2CO_3 and under oxidizing conditions are retained as $MgSO_4$ and Na_2SO_4. The sulfate in the residue is extracted and determined gravimetrically.

In the bomb washing method, sulfur is determined in the washings from the oxygen bomb calorimeter following the calorimetric determination (method D 2015 or D 3286). After opening, the inside of the bomb is washed carefully, and the washings are collected. After titration with standard base solution to determine the acid correction for the heating value, the solution is heated and treated with NH_4OH to precipitate iron ions as $Fe(OH)_3$. After filtering and heating, the sulfate is precipitated with $BaCl_2$ and determined gravimetrically.

In the high-temperature combustion methods, a weighed sample is burned in a tube furnace at a minimum operating temperature of 1,350°C in a stream of oxygen to ensure the complete oxidation of sulfur-containing components in the sample. Using these conditions, all sulfur-containing materials in the coal or coke are converted predominantly to sulfur dioxide in a reproducible way. The amount of sulfur dioxide produced by burning the sample can be determined by the three alternative methods mentioned previously. Diagrams of the apparatus used in each of the three methods are given in Figures 5.3, 5.4, and 5.5.

In the acid-base titration method, the combustion gases are bubbled through a hydrogen peroxide solution in a gas absorption bulb. Sulfuric acid is produced by the reaction of sulfur dioxide with the hydrogen peroxide and is determined by titration with a standard base solution. Chlorine-containing species in the sample yield hydrochloric acid in the hydrogen peroxide solution, which contributes to its total acidity. For accurate results, a correction must be made for the chlorine present in the sample. This can be done by

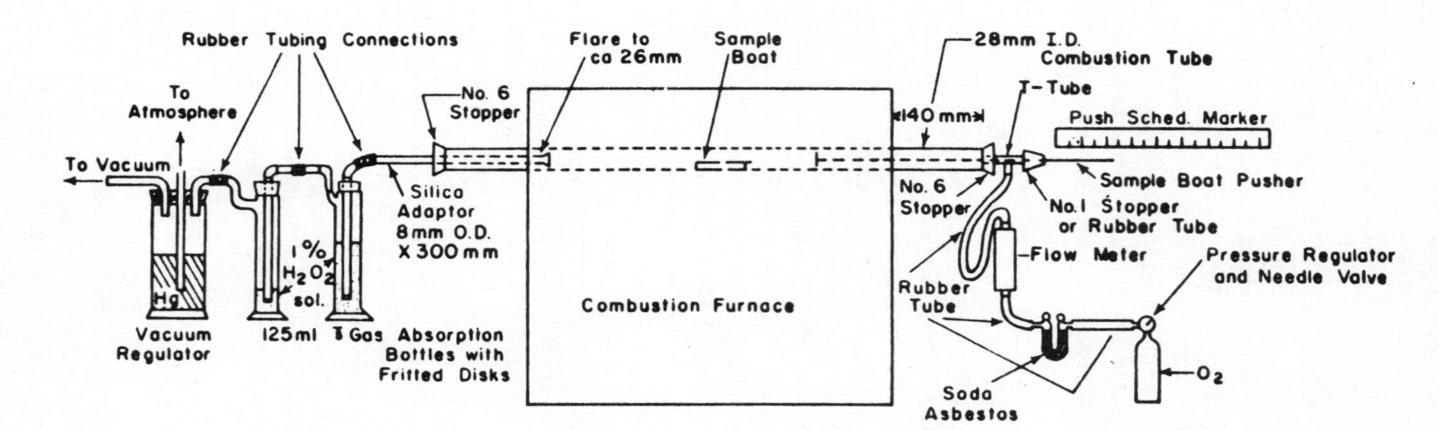

FIGURE 5.3 Apparatus for the determination of sulfur using acid-base titration. (*Source:* Reprinted with permission from *Annual Book of ASTM Standards,* vol. 05.05, copyright ASTM, 1916 Race St., Philadelphia, PA 19103.)

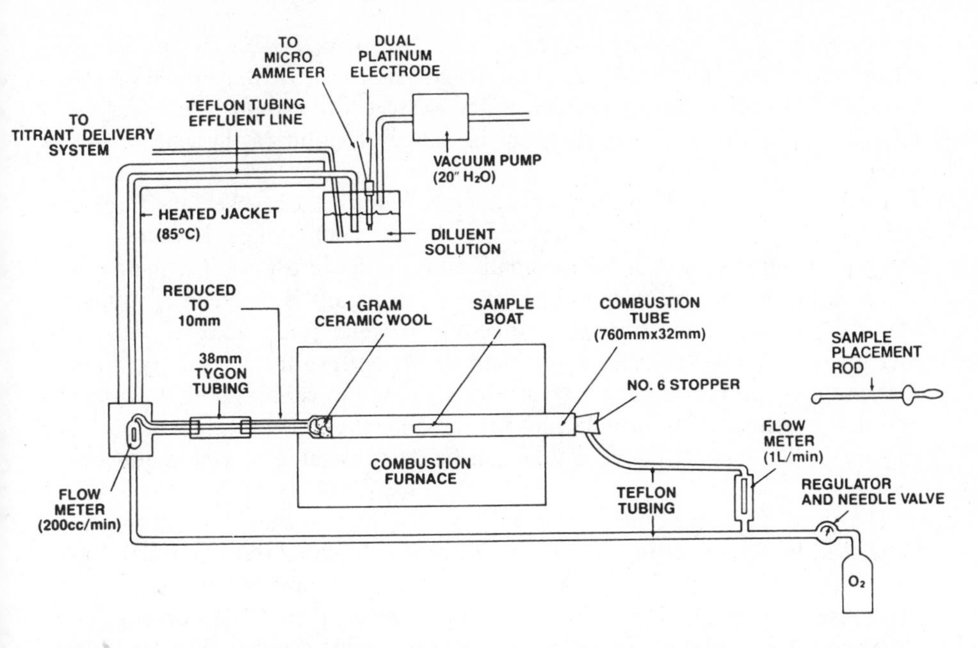

FIGURE 5.4 Apparatus for the determination of sulfur by the iodimetric detection method. (*Source:* Reprinted with permission from *Annual Book of ASTM Standards,* vol. 05.05, copyright ASTM, 1916 Race St., Philadelphia, PA 19103.)

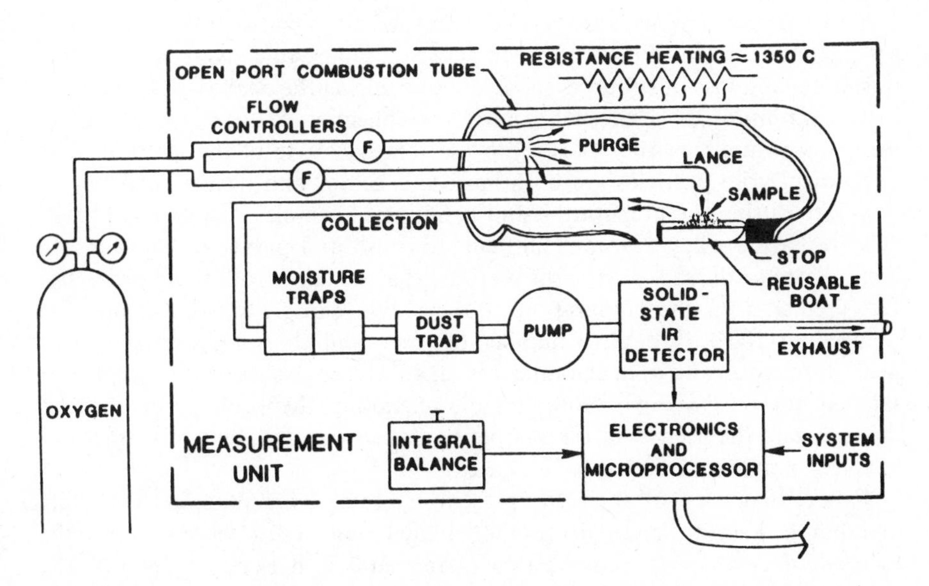

FIGURE 5.5 Apparatus for the determination of sulfur by the infrared detection method. (*Source:* Reprinted with permission from *Annual Book of ASTM Standards,* vol. 05.05, copyright ASTM, 1916 Race St., Philadelphia, PA 19103.)

determining the chlorine by ASTM method D 2361 or D 4208, which are discussed later. Appropriate standard reference materials (SRMs) should be used to calibrate commercially available sulfur analyzers to establish recovery factors or a calibration curve based on the range of sulfur in the coal or coke samples being analyzed.

In the iodimetric titration procedure, the combustion gases are bubbled through a diluent solution containing pyridine, methanol, and water. This solution is titrated with a titrant containing iodine in a pyridine, methanol, and water solution. In automated systems, the titrant is delivered automatically from a calibrated burette syringe and the end point detected amperometrically. The method is empirical, and SRMs with sulfur percentages in the range of the samples to be analyzed should be used to calibrate the instrument before use. Alternative formulations for the diluent and titrant may be used in this method to the extent that they can be demonstrated to yield equivalent results.

The third method of measuring the sulfur dioxide in the combustion gases is by the absorption of infrared (IR) radiation. Moisture and particulates are first removed from the gas stream by traps filled with anhydrous magnesium perchlorate. The gas stream is then passed through an IR absorption cell tuned to a frequency of radiation absorbed by sulfur dioxide. The IR radiation absorbed during combustion of the sample is proportional to the sulfur dioxide in the combustion gases and therefore to the sulfur in the sample. The method is empirical, and SRMs with sulfur percentages in the range of the samples to be analyzed should be used to calibrate the instrument before use.

Some general problems associated with the determination of sulfur in coal are nonuniform distribution of pyrite particles, failure to convert all the sulfur to sulfate, and loss of sulfur as sulfur dioxide during the analysis. The nonuniform distribution of pyrite necessitates the collection of many sample increments to ensure that the gross sample is representative of the lot of coal in question. Pyrite particles are both hard and heavy, and have a tendency to segregate during the preparation and handling of samples. Because the particles are harder, they are more difficult to crush and pulverize, and tend to concentrate in the last portion of material that remains from these processes. Requirements for the preparation of an analysis sample are that it must be reduced to 100% through a number 60 sieve and thoroughly mixed. This procedure ensures that, at the time it is taken, the analysis sample is representative of the gross sample. However, upon handling, the heavy pyrite particles do segregate themselves in the sample bottle, so that frequent mixing of the analysis sample is necessary as it is used.

Failure to convert all the sulfur present in a sample to sulfate in the Eschka and bomb washing methods results in low total sulfur values. Since the methods depend on the combustion of the sample, it is important that the combustion products are completely oxidized. The temperature used must be high enough, the rate of burning must not be too fast, and sufficient time must be allowed to complete the conversion.

In each of the methods discussed, sulfur is oxidized to sulfur dioxide during the analysis. Some SO_2 may be lost unless the necessary precautions are taken. In the Eschka method, a generous layer of Eschka mixture covering the fusion mixture helps prevent the loss of sulfur as sulfur dioxide. The mixture must be heated gradually to guard against the production of SO_2 at a rate that is too high for it to be absorbed by the Eschka mixture. In the bomb washing method, the pressure of the bomb should be released slowly after the sample is burned in oxygen so that sulfur oxides will not be carried out of the bomb. In the high-temperature combustion methods, it is essential that the flow of oxygen is sufficient and that the rate of heating is not too high. A high rate of heating will lead to the evolution of combustion products, including sulfur dioxide, at a rate that is too rapid for complete absorption in the solutions or for detection by the IR cell.

The gravimetric determination of sulfate can be and is most often used to finish the Eschka and bomb-washing methods. The most serious problem that arises concerns the barium sulfate precipitate. It may be extremely fine and difficult to filter. One way to obtain a $BaSO_4$ precipitate that is easily filtered is to add the $BaCl_2$ precipitant rapidly to the hot solution and stir the mixture vigorously. Heating and digestion for a lengthy period improve the filterability of the precipitate. Addition of a slurry of filter paper, prepared by digesting small pieces of paper in hot water, acts as a filter aid. After filtering, the precipitate must be washed several times with hot water to remove adsorbed materials that will cause the results to be too high. Barium sulfate is a rather strong adsorbing agent and readily adsorbs iron during the precipitation. Whenever the iron content of the coal sample is high, the iron should be removed through precipitation and filtering before the sulfate is precipitated.

Interpretation and Uses of Total Sulfur Data. Total sulfur data are necessary for the effective control of the emissions of oxides of sulfur whenever coal is used as a fuel. The emission of sulfur oxides leads to the corrosion of equipment and slagging of combustion or boiler equipment, as well as contributing to atmospheric pollution and environmental damage. Sulfur data are therefore necessary for the evaluation of coals to be used for combustion purposes.

Most coal conversion and cleaning processes require two sets of sulfur values: the sulfur content of the coal before it is used and the sulfur content of the products formed. In the coking of coal, some of the sulfur is removed in the coking process, which makes it necessary to obtain the before and after values. The commercial uses of coke, as in metallurgical processes, require a low sulfur content and necessitate an accurate sulfur value for the coke. In coal gasification and liquefaction processes, the sulfur in the coal is sometimes carried through to the products. It is therefore necessary to determine the amount of sulfur in each of the products before it is used. One of the primary reasons for cleaning coal is to reduce the sulfur content. It is necessary to know the sulfur content before and after cleaning in order to evaluate the cleaning process.

Total sulfur values alone are not adequate in accessing a cleaning process for reducing the sulfur content of coal. Pyritic sulfur alone can be removed by specific gravity separations, and its removal depends on the way the pyrite is distributed throughout the coal. If it occurs as very small crystals widely dispersed in the coal, it is almost impossible to remove by these methods. When pyrite occurs in large pieces, it can be successfully removed by specific gravity methods. Organic sulfur cannot be reduced appreciably, since it is usually uniformly dispersed throughout the organic material in coal.

5.4.4 Oxygen

Oxygen occurs in both the organic and inorganic portions of coal. In the organic portion, oxygen is present in hydroxyl, carboxyl, methoxyl, and carbonyl groups. In low-rank coals, the hydroxyl oxygen averages about 6–9%, while high-rank coals contain less than 1%. The percentages of oxygen in carbonyl, methoxyl, and carboxyl groups average from a few percent in low-rank and brown coals to almost no measurable value in high-rank coals (12).

The inorganic materials in coal that contain oxygen are the various forms of moisture, silicates, carbonates, oxides, and sulfates. The silicates are primarily aluminum silicates found in the shalelike portions. Most of the carbonate is calcium carbonate, the oxides are mainly iron oxides, and the sulfates are calcium and iron sulfates.

Determination of Oxygen Content. Currently there is no satisfactory direct ASTM method of determining oxygen content. In an ultimate analysis, it is calculated by subtracting the sum of the as-determined percentages of C, H, N, S, and ash from 100. This estimated value is affected by errors incurred in the determinations of the values for the other elements and by changes in the weight of the ash-forming constituents on ignition. The oxygen value calculated as a weight percentage of the analysis sample according to this procedure does not include the oxygen in the ash but does include the oxygen in the moisture associated with the analysis sample.

The most widely used direct method of determining oxygen in coal is known as the Schütze-Unterzaucher method, with modifications by a variety of workers (13–15). Generally, the procedure is to pyrolyze the coal in a stream of dry nitrogen. The volatilized products are passed over carbon at about 1,100°C (900°C using a platinum-carbon catalyst), which converts the oxygen in the volatile products to carbon monoxide. The carbon monoxide is then oxidized to carbon dioxide, usually with iodine pentoxide, which releases free iodine. The iodine released can be determined titrimetrically, or the carbon dioxide produced can be absorbed and determined gravimetrically to calculate the amount of oxygen in the original samples.

The basic principles involved in the Schütze-Unterzaucher method of determining the oxygen content of coal may lead one to believe it is relatively simple. However, the method is a complicated one, is time consuming, requires special equipment and reagents, and has many other problems associated with its use. The varied sources of oxygen in coal—such as the oxygen in the moisture and water of hydration of mineral matter, the oxygen in carbonates, and the oxygen in silicates and other inorganic compounds in addition to the oxygen in the organic matter—all offer difficulties. The original procedure has been modified in several ways to reduce the contribution made by some of these oxygen sources to the determined oxygen value. Thorough drying in a nitrogen atmosphere before the pyrolysis of the sample minimizes the effect of moisture, and much of the mineral matter is removed by a specific gravity separation or chemical treatment with hydrochloric and hydrofluoric acid. The reduction of mineral matter minimizes the contribution that the water of hydration and the inorganic compounds, such as carbonates, silicates, oxides, and sulfates, make to the determined oxygen value. The oxygen value obtained by this method, after all the pretreatment steps are taken to remove moisture and mineral matter, is essentially a measure of the oxygen contained in the organic matter in coal.

The precision and accuracy obtained in using the direct method of determining oxygen content is not as reliable as that obtained in other analytical methods used in coal analysis. However, the direct method does allow one to obtain a more precise value of the oxygen content of coal then can be obtained in the determination of oxygen content by the difference method.

Interpretation and Uses of Oxygen Data. When the oxygen value is estimated by subtracting the determined percentages of all other constituents from 100, the errors in the determined values are reflected in the estimated oxygen value. These errors may be partially compensating, or they may be additive. It is important that accurate determinations are made and appropriate corrections for overlapping values, especially hydrogen, be calculated.

The following is an expression for calculating the percent oxygen according to the ASTM definition:

$$O_x = 100 - [C + H + N + S + \text{ash}] \tag{4}$$

where all values are expressed as percentages and C = as-determined carbon (method D 3178), H = as-determined hydrogen (method D 3178), N = as-determined nitrogen (method D 3179), S = as-determined sulfur (method D 3177 or D 4239), and ash = as-determined ash (method D 3174). All these above values pertain to those obtained from the analysis sample. The as-determined value for carbon represents both the organic and the carbonate carbon. The as-determined value for hydrogen represents the organic hydrogen, the hydrogen in the residual moisture, and the hydrogen in the water of hydration of mineral matter. The sulfur value is the total sulfur in coal and represents that which is contained in the organic matter, pyrites, and sulfates. Ash is mostly metal and silicon oxides. The estimated value of the oxygen therefore includes the oxygen contained in the organic matter, in the moisture, and in the mineral matter, except that which is combined with metals in coal ash.

A rough estimate of the oxygen contained in the organic matter in coal can be obtained by correcting the oxygen value obtained by equation 4 for the oxygen in residual moisture and water of hydration of mineral matter. Adding this correction to the sum subtracted from 100 gives the following expression:

$$O_x = 100 - [C + H + N + S + \text{ash} + 8/9(H_2O + H_2O \text{ of hydration})] \quad (5)$$

where H_2O = as-determined residual moisture (ASTM Method 3173) and H_2O of hydration = 8.0% of ash.

Several improvements in the estimation of organic oxygen can be made when the analytical data are available. Values for chlorine, carbon dioxide, pyritic sulfur, and sulfur in coal ash are helpful in improving the estimation. Failure to include the chlorine value in the sum subtracted from 100 according to the equation 4 leads to a high value for the organic oxygen. This oxygen value should also be reduced for the oxygen present in the CO_2 that is associated with the mineral matter. The inclusion of both total sulfur and ash in the sum that is subtracted from 100% in estimating the oxygen content of coal lowers the oxygen value, since part of the sulfur may be retained in the ash. The sulfur that is retained is therefore counted twice in the sum for subtraction. Correcting the coal ash for the SO_3 present compensates for this error. Likewise, the coal ash should be corrected for any Fe_2O_3 that results from the heating of pyrite (FeS_2) in air, as is done in the ashing process. In the ashing process, 3 oxygen atoms replace 4 sulfur atoms, as is illustrated by the equation

$$4FeS_2 + 11O_2 \longrightarrow 2Fe_2O_3 + 8SO_2$$

On a weight basis, 48 parts of oxygen replace 128 parts of sulfur. This oxygen is from an external source and not from the coal itself. Since this oxygen contributes to the weight of the ash, a correction of $\frac{3}{8}$ of the pyritic sulfur value is necessary. The pyritic sulfur that is replaced is accounted for in the total sulfur value. When the values are available to make these corrections, a good estimate of the percentage of oxygen in the organic or combustible portion of coal can be made, according to the following formula:

$$\begin{aligned} O_x = 100 - [&C + H + N + S + Cl + (\text{ash} - 3/8 S_p - SO_3 \text{ in ash}) \\ &+ 8/9(H_2O + H_2O \text{ of hydration}) + 32/44\, CO_2] \end{aligned} \quad (6)$$

where Cl = as-determined Cl (method D 2361 or D 4208), S_p = as-determined pyritic sulfur (method D 2492), SO_3 in ash = as-determined SO_3 in ash (method D 1757), and CO_2 = as-determined carbon dioxide in coal (method D 1756). All other terms are as given in the previous formulas, and all values are expressed as percentages.

In the uses and applications of oxygen data, the most important value is the oxygen content of the organic matter in coal. This can be estimated by one of the above formulas, depending on the information available. If the oxygen is determined directly, using the method previously discussed in which the moisture and mineral matter are removed from the sample before the actual oxygen determination, the oxygen value in this case represents that contained in the organic matter. In calculating heat balances for boiler efficiency studies, it is important that an accurate value of the combustible material in coal be obtained. Thus, a correction for the oxygen content of the organic matter of coal should be made. Of course, corrections to the carbon and hydrogen values for the amount of these elements found in the moisture and inorganic constituents of coal should also be made. Oxygen data are used for determining the suitability of coals for coking, liquefaction, or gasification processes. In general, coals with a high oxygen content are unsuitable for coking but may be more reactive and thus easier to gasify or liquefy.

5.5 CALCULATING COAL ANALYSES FROM AS-DETERMINED VALUES TO DIFFERENT BASES

The results of a coal analysis may be reported on any of a number of bases, differing from each other in the manner by which moisture is treated. Except for data reported on a dry basis, it is essential that an appropriate moisture content be given in the data report. This would avoid ambiguity and provide a means for conversion of data to other bases. These bases are the following:

As-determined basis (ad): the basis for analytical data obtained from an analysis sample of coal or coke prepared and reduced to 100% through a number 60 (250 μm) sieve, according to method D 2013; represents the values obtained at the particular moisture level in the analysis sample at the time of the analysis

As-received basis (ar): the basis for analytical data calculated to the moisture condition of the sample as it arrived at the laboratory and before any processing is begun; represents the moisture basis for the sample as it was taken, provided it was properly sealed so there was no gain or loss of moisture

Dry basis (d): the basis for analytical data calculated to a theoretical basis of no moisture associated with the sample, with the moisture value as determined in method D 3173 used for converting the as-determined data to a dry basis

Dry, ash-free basis (daf): the basis for data calculated to a theoretical basis of no moisture or ash associated with the sample, with values obtained by methods D 3173 and D 3174 used for converting the as-determined data to this basis

Equilibrium moisture basis: the basis for data calculated to the moisture level established as the equilibrium moisture, with values obtained as in method D 1412 used to make this conversion

5.5.1 Formulas for Converting Data

In converting from the as-determined (*ad*) basis to the as-received (*ar*) basis, the following formulas are used (all values are expressed in weight percent).

For moisture (*M*),

$$M_{ar} = M_{ad} \times \frac{100 - ADL}{100} + ADL \tag{7}$$

where ADL = air-dry loss in weight percent of as-received sample (see equation 1).

For hydrogen (H) and oxygen (O), including the hydrogen and oxygen in the moisture associated with the sample,

$$\mathrm{H}_{ar} = (\mathrm{H}_{ad} - 0.1119\ M_{ad}) \times \frac{100 - ADL}{100} + 0.1119\ M_{ar} \tag{8}$$

$$\mathrm{O}_{ar} = (\mathrm{O}_{ad} - 0.8881\ M_{ad}) \times \frac{100 - ADL}{100} + 0.8881\ M_{ar} \tag{9}$$

For hydrogen and oxygen not including the hydrogen and oxygen in the moisture associated with the sample,

$$\mathrm{H}_{ar} = (\mathrm{H}_{ad} - 0.1119\ M_{ad}) \times \frac{100 - ADL}{100} \tag{10}$$

$$\mathrm{O}_{ar} = (\mathrm{O}_{ad} - 0.8881\ M_{ad}) \times \frac{100 - ADL}{100} \tag{11}$$

In converting from the as-determined to the dry (d) basis, the following formulas apply to hydrogen and oxygen:

$$\mathrm{H}_{d} = (\mathrm{H}_{ad} - 0.1119\ M_{ad}) \times \frac{100}{100 - M_{ad}} \tag{12}$$

$$\mathrm{O}_{d} = (\mathrm{O}_{ad} - 0.8881\ M_{ad}) \times \frac{100}{100 - M_{ad}} \tag{13}$$

In converting all other parameters from one basis to another, the following general formula applies when using the appropriate conversion factor from Table 5.4:

$$\mathrm{P}_{\text{wanted}} = \mathrm{P}_{\text{given}} \times \text{conversion factor} \tag{14}$$

TABLE 5.4 **Conversion Factor Chart**

	Wanted			
Given	As-Determined (ad)	As-Received (ar)	Dry (d)	Dry, Ash-Free (daf)
As-determined (ad)	—	$\frac{100 - ADL}{100}$	$\frac{100}{100 - M_{ad}}$	$\frac{100}{100 - M_{ad} - A_{ad}}$
As-received (ar)	$\frac{100}{100 - ADL}$	—	$\frac{100}{100 - M_{ar}}$	$\frac{100}{100 - M_{ar} - A_{ar}}$
Dry (d)	$\frac{100 - M_{ad}}{100}$	$\frac{100 - M_{ar}}{100}$	—	$\frac{100}{100 - A_d}$
Dry, ash-free (daf)	$\frac{100 - M_{ad} - A_{ad}}{100}$	$\frac{100 - M_{ar} - A_{ar}}{100}$	$\frac{100 - A_d}{100}$	—

The parameters to which equation 13 and Table 5.4 apply are the following:

Ash (A)	Fixed carbon
Calorific value (gross)	Nitrogen
Carbon dioxide	Sulfur
Chlorine	Sulfur forms (pyritic, sulfate, and organic)
	Volatile matter

The parameters must be expressed as a weight percentage, except gross calorific value, which is expressed as Btu/lb. An example of ultimate analysis data that has been calculated to various bases is given in Table 5.5.

Additional information and formulas for converting data to other bases can be found in ASTM method D 3180 or in the U.S. Bureau of Mines publication *Methods of Analyzing and Testing Coal and Coke* (16).

5.6 MISCELLANEOUS ANALYSIS

The category of miscellaneous analysis encompasses element-related analyses—such as the determination of chlorine, forms of sulfur, carbon dioxide, and trace elements in coal—as well as special tests—such as the determination of the calorific value, fusibility of coal and coke ash, free-swelling index, grind-

TABLE 5.5 Ultimate Analysis Data

Parameter[a]	As-Determined[b]	Dry	As-Received[b]	As-Received[c]
Carbon	68.30	73.21	64.48	64.48
Hydrogen	5.49	5.08	5.81	4.48
Nitrogen	1.19	1.28	1.12	1.12
Sulfur	2.60	2.79	2.45	2.45
Ash	7.11	7.62	6.71	6.71
Oxygen	15.31	10.02	19.43	8.84
Total %	100.00	100.00	100.00	88.08
Moisture				
Air-dry loss moisture	5.60			
Moisture (analysis sample)	6.70			
Total moisture				11.92
Total %				100.00

[a] All values are given in percent by weight.
[b] Hydrogen and oxygen include H and O in sample moisture (M_{ar}).
[c] Hydrogen and oxygen do not include H and O in sample moisture (M_{ar}).

ability, and plastic properties of coal. Other tests can be included in this category, but the ones listed are those that are discussed here.

5.6.1 Chlorine

The chlorine content of coal is normally low, usually only a few tenths of a percent or less. It occurs predominantly as sodium, potassium, and calcium chlorides, with magnesium and iron chlorides present in some coals. There is evidence that chlorine may also be combined with the organic matter in coal (17).

Determination of Chlorine Content. Methods of converting the chlorine in coal into a form suitable for its analytical determination include combusting the sample, with or without Eschka mixture, in an oxygen bomb, and heating with Eschka mixture in an oxidizing atmosphere. Eschka mixture is a combination of two parts by weight of magnesium oxide and one part of anhydrous sodium carbonate. There are two standard methods of determining chlorine in coal: method D 2361 and method D 4208.

ASTM method D 2361 offers a choice of two procedures for combusting the coal sample. In the bomb combustion procedure, the oxygen bomb used is the same as, or very similar to, that used in the determination of the calorific value as described in ASTM methods D 2015 and D 3286. In the determination, 1 g of the analysis sample of coal is mixed with 1 g of Eschka mixture and placed in a crucible inside an oxygen bomb. An ammonium carbonate solution is added to the bomb to trap the chloride containing species produced in the combustion. After charging with oxygen to 25 atmospheres, the bomb is fired and allowed to stand in the calorimeter water for at least 10 min. The pressure on the bomb is then released slowly, the bomb is disassembled, and all parts of the bomb interior are washed with hot water. The washings are collected in a beaker and acidified with nitric acid. The amount of chloride in the solution is then determined by either a modified Volhard or a potentiometric titration with silver nitrate solution.

In the second procedure of ASTM method D 2361, 1 g of the coal analysis sample is mixed with 3 g of Eschka mixture in a suitable crucible. This mixture is covered with an additional 2 g of Eschka mixture to ensure that no chlorine is lost during combustion. The mixture is then ignited gradually in a muffle furnace by raising the temperature to 675° C ± 25° C within 1 h. This temperature is maintained for $1\frac{1}{2}$ h before the cooling and washing of the incinerated mixture with hot water into a beaker. The contents of the beaker are acidified with nitric acid, and the chloride is determined as in the previously described procedure.

In ASTM method D 4208, 1 g of the analysis sample of coal is placed in a crucible inside an oxygen bomb. A sodium carbonate solution is added to the bomb to trap the chloride containing species produced. After charging with oxygen to 25 atmospheres, the bomb is fired and allowed to stand in the calorimeter water for at least 15 min. After the pressure is released very slowly, the bomb is disassembled, and all parts of the bomb interior are washed with water. The washings are collected, an ionic strength adjuster ($NaNO_3$) is added, and the chloride is determined with an ion-selective electrode by the standard addition method.

In both methods, it is possible to lose some of the chlorine during combustion unless necessary precautions are taken. Thoroughly mixing the coal sample with Eschka mixture and carefully covering this with additional Eschka mixture minimize the loss of chlorine. In the bomb combustion methods, the ammonium and sodium carbonate solutions in the bomb are used to absorb the chlorine as it is released in the combustion. The 10-and 15-min waiting periods and the slow release of the pressure on the bomb help to prevent the loss of chlorine as well.

Interpretation and Uses of Chlorine Data. The chlorine in coal and in the products derived from coal is known to contribute significantly to the corrosion of the coal handling and processing equipment. Since the corrosion of this equipment is the result of several causes, one being the chlorine content of coal, it is difficult to predict the degree of corrosion within a given time frame. It is equally as difficult to predict the degree to which the chlorine content contributes to the corrosion, other than the general prediction that the higher the chlorine content, the greater the chances for corrosion of the equipment. As a general rule, coals with a high chlorine content are not desirable.

Chlorine data are used in ultimate analysis to improve the estimate of oxygen by difference. The chlorine value is included in the sum of the items determined, which, when subtracted from 100, gives an estimate of the oxygen content of coal.

5.6.2 Forms of Sulfur

Sulfur occurs in coal as inorganic sulfates, as pyrites, and in combination with the organic matter. Organic sulfur and pyrites account for almost all the sulfur in coals. Sulfate sulfur is usually less than 0.1%, except for in weathered coal containing an appreciable amount of pyrites. The pyritic sulfur content varies considerably more than does the organic sulfur content and is of more interest because it is the form that can be most easily removed from coal by current preparation practices.

Determination of the Content of the Forms of Sulfur. The procedures for determining the forms of sulfur in coal are described in ASTM method D 2492. In this method, the sulfate and pyrite sulfur are determined directly, and the organic sulfur is taken as the difference between the total sulfur and the sum of the sulfate and pyritic sulfur.

In the determination of sulfate, 2–5 g of the analysis sample are mixed with HCl (2 volumes concentrated HCl + 3 volumes of water), and the mixture is gently boiled for 30 min. After filtering and washing, the undissolved coal may be retained for the determination of pyritic sulfur, or it may be discarded and a fresh sample used for pyritic sulfur. Saturated bromine water is added to the filtrate to oxidize all sulfur forms to sulfate ions and ferrous ions to ferric ions. After boiling to remove excess bromine, the iron is precipitated with excess ammonia and filtered. This precipitate must be retained for the determination of nonpyritic iron if a fresh sample of coal was used for the determination of the pyritic iron. The sulfate is then precipitated with $BaCl_2$, and the $BaSO_4$ is determined gravimetrically.

Either the residue from the sulfate determination or a fresh 1-g sample is used for the determination of pyritic sulfur content. The sample is added to dilute nitric acid and the mixture boiled gently for 30 min or allowed to stand overnight. This treatment oxidizes iron species to iron(III) and inorganic sulfur compounds to sulfate. The mixture is then filtered, and the filtrate is saved for the determination of iron by atomic absorption spectrophotometry or by a titration procedure. If iron is to be determined by the atomic absorption method, no further work is done on the filtrate other than to dilute it to an appropriate volume before the determination. If a titration method is to be used for the determination of iron, the filtrate is treated with 30% H_2O_2 to destroy any coloration arising from the coal. The iron is then precipitated, filtered, and washed. The precipitate is then dissolved in HCl, and the iron determined by titration with either $K_2Cr_2O_7$ or $KMnO_4$.

If a new sample was used for the determination of the pyritic iron, the iron determined by these procedures represents the combination of the pyritic and nonpyritic iron. The amount of nonpyritic iron must then be determined separately and subtracted from the amount determined by the methods described here. If the residue from the sulfate determination was used, the iron determined by the above procedures represents the pyritic iron. Once the correct value for the pyritic iron is determined, the pyritic sulfur is calculated using the following expression:

$$\%\ \text{Pyritic sulfur} = \%\ \text{pyritic iron} \times 2 \times 32.06/55.85$$

where $2 \times 32.06/55.85$ is the ratio of sulfur to iron in pyrite.

Some difficulties encountered in determining the amounts of the various forms of sulfur in coal are adsorption of other materials on $BaSO_4$ when it is precipitated, inability to extract all the pyritic sulfur from the coal during the extraction process, and possible oxidation of pyritic sulfur to sulfate in the pulverization and storage of the coal sample. Both the adsorption of other materials on $BaSO_4$ and oxidation of pyritic sulfur lead to high values for the sulfate sulfur. Iron ions are readily adsorbed on $BaSO_4$, which could be particularly objectionable for coals containing large amounts of nonpyritic iron. Removal of the iron by precipitation and filtration before the precipitation of

$BaSO_4$ minimizes the adsorption of the iron. Inadequate pulverization and mixing of the sample appear to be the major causes of the incomplete extraction of pyritic sulfur from coal. A very small amount of organic sulfur may also be extracted with the pyritic sulfur. For this reason, the amount of pyritic iron extracted is used as a measure of the pyritic sulfur. To control the oxidation of pyritic sulfur to sulfates, exposure of the coal sample to the atmosphere at elevated temperatures should be avoided, and the sample should be analyzed as soon as possible. A discussion of other problems associated with the determination of sulfur as well as efforts to develop new methods of analysis has been given by J. K. Kuhn (18).

Interpretation and Uses of Forms of Sulfur Data. The principal use of forms of sulfur data is in connection with the cleaning of coal. Within certain limits, pyritic sulfur can be removed from coal by gravity separation methods, whereas organic sulfur cannot. Pyritic sulfur content can therefore be used to predict how much sulfur can be removed from the coal and to evaluate cleaning processes. If the pyritic sulfur occurs in layers, it usually can be removed efficiently. If it occurs as fine crystals dispersed throughout the coal, its removal is very difficult.

Other uses of forms of sulfur data are the inclusion of the pyritic sulfur value in the formula for the estimation of oxygen by difference and as a possible means of predicting the extent of weathering of coal. The sulfate concentration increases upon weathering, so the sulfate sulfur value could be used as an indication of the extent of weathering of coal.

5.6.3 Carbon Dioxide in Coal

Most coals contain small amounts of mineral carbonates made up primarily of calcium carbonate and to a lesser extent ferrous and other metal carbonates. Some coals contain a comparatively large amount of the inorganic carbonates, and the determination of carbon dioxide content is required in estimating the mineral matter content of these high-carbonate coals.

Determination of Carbon Dioxide Content. In summary, the determination of the carbon dioxide content of coal is made by decomposing, in a closed system, a weighed sample of coal with hydrochloric acid, which liberates the carbon dioxide. This is absorbed in a CO_2 absorbent, such as NaOH or KOH on an inert carrier. The increase in mass of the absorbent is a measure of the CO_2 released by the coal sample, which can be used to calculate the amount of mineral carbonates in the coal.

Due to the small amount of carbon dioxide in coal and the difficulty of accurately measuring the CO_2 that is liberated, some strict requirements have been set for the construction and design of the apparatus to be used. These requirements are given in detail in ASTM method D 1756. The apparatus must contain an air flow meter and purifying train, a reaction unit fitted with a separatory funnel and water-cooled condenser, a unit for removing interfering gases, and an absorber. The air-purifying train removes all carbon dioxide, and the water-cooled condenser removes moisture before it can enter the absorption train. Acid-forming gases, such as SO_2, H_2S, and halogen acids are produced in the reaction and must be removed before entering the CO_2 absorber. Otherwise, they will be weighed as absorbed and measured as carbon dioxide. Anhydrous copper sulfate on pumice or granular silver sulfate is positioned in the absorption train to remove these interfering gases from the air stream before it enters the CO_2 absorber. The entire system must be gas-tight in order to prevent error, and a time schedule is specified to ensure repeatability and reproducibility.

Interpretation and Uses of Carbon Dioxide Data. The carbon dioxide value is used primarily in the estimation of the mineral matter of high-carbonate coals. When the carbon dioxide value is high, it is also used to correct volatile matter values. A high value indicates a large amount of calcium carbonate, which retains sulfur as sulfate quite readily during combustion. Consequently, this gives a high ash value as well.

5.6.4 Calorific Value of Coal

The calorific value of a coal is the combined heats of combustion of the carbon, hydrogen, nitrogen, and sulfur in the organic matter and of the sulfur in pyrite. The energy released upon combustion is of primary interest to coal producers and users. The calorific value, on a specified basis, is one of the more important parameters employed in the classification of coals.

Determination of the Gross Calorific Value of Coal. The common method of determining the gross calorific value of coal is with either an adiabatic or an isothermal bomb calorimeter. The procedures for using these calorimeters are specified in ASTM methods D 2015 and D 3286. In these methods, a weighed sample is burned in an oxygen bomb covered with water in a container surrounded by a jacket. In an adiabatic calorimeter system (method D 2015), the jacket temperature is adjusted during the burning so that it is essentially the same as the calorimeter water temperature. In an isothermal calorimeter system (method D 3286), the temperature rise of the calorimeter water is corrected for the heat lost to or gained from the surrounding jacket during the burning of the sample. In both systems, the corrected temperature rise times the energy equivalent of the calorimeter gives the total amount of heat produced during the burning of the sample. The energy equivalent (also called the water equivalent or heat capacity) of the calorimeter is determined by burning standard samples of benzoic acid.

After firing, the contents of the bomb are washed into a beaker and titrated with standard sodium carbonate solution to determine the amount of acid (HNO_3 and H_2SO_4) produced in the combustion. Corrections for the amount of acid, the amount of fuse wire used in firing, and the sulfur content of the sample are then made to the total heat produced in the calorimeter (energy equivalent times corrected temperature rise) to determine the gross calorific value of the solid fuel.

Perhaps the greatest source of error in this method is in temperature measurement. If mercury in glass thermometers are used, they must be calibrated well, and consistent readings must be made. Many calorimeters are equipped with digital thermometers with thermistor probes and microprocessors to control the firing and record the temperatures at prescribed intervals. This alleviates most of the human error in recording the temperature changes.

Igniting the coal sample in the oxygen bomb can be difficult. The sample may be blown out of the crucible by introducing the oxygen too quickly. Pressing the coal sample into a pellet may prevent the sample from blowing out. Coals with a high mineral content are hard to ignite, and mixing the sample with a measured amount of standard benzoic acid and pelleting may be helpful.

After firing, restoring the bomb pressure to atmospheric pressure too rapidly may result in the loss of oxides of sulfur and nitrogen. A correction must be made to the gross calorific value for the amounts of these acid forming oxides produced in the bomb. Their loss results in a high calorific value. The pressure of the bomb must be restored very slowly to prevent this.

The equipment used must be checked periodically for any changes in the energy equivalent of the calorimeter, any corrosion or damage to the calorimeter bucket, any damage (however slight it may be) to the oxygen bomb, and any malfunction of the stirrers, electrical system, or other parts of the calorimeter. Any of these changes or malfunctions may change the energy equivalent of the calorimeter or introduce extra heat, which would lead to errors in the measured calorific value.

Interpretation and Uses of Calorific Data. The calorific value is normally the basic item specified in contracts for coal to be used in steam plants. It is the most important value determined for coal that is to be used for heating purposes. In coal contracts, the calorific value is usually specified on the as-received basis. Any error in the moisture value is reflected in the as-received calorific value.

The laboratory-determined calorific value is called the gross calorific value. It may be defined as the heat produced by combustion of a unit quantity of coal at constant volume in an oxygen bomb calorimeter under specified conditions such that the end products of the combustion are in the form of ash, gaseous carbon dioxide, sulfur dioxide, nitrogen, and liquid water. Burning coal as a fuel does not produce as much heat per unit quantity. Corrections are made to the gross calorific values for this difference between the laboratory and coal-burning facility. The corrected value is referred to as the net calorific value. This is defined as the heat produced by combustion of a unit quantity of coal at constant atmospheric pressure under conditions such that all water in the products remains in the form of vapor. The net calorific value is lower than the gross calorific value.

In the ASTM system of classifying coals by rank, the calorific value is used as one of the main parameters for the classification of bituminous, subbituminous, and lignitic coals. Coal calorific values are also used in estimating resources.

5.6.5 Fusibility of Coal Ash

Coal ash is the noncombustible residue that remains after all the combustible material has been burned. It is a complex mixture that results from chemical changes that take place in the components of the coal mineral matter. The composition of coal ash varies extensively just as the composition of coal mineral matter varies.

The ash fusibility determination is an empirical test designed to simulate as closely as possible the behavior of coal ash when it is heated in contact with either a reducing or an oxidizing atmosphere. The test is intended to provide information on the fusion characteristics of the ash. It gives an approximation of the temperatures at which the ash remaining after the combustion of coal will sinter, melt, and flow. Sintering is the process by which the solid ash particles weld together without melting. The temperature points are measured by observation of the behavior of triangular pyriamids (cones) prepared from coal ash when heated at a specified rate in a controlled atmosphere. The critical temperature points are as follows:

Initial deformation temperature (IT): temperature at which the first rounding of the apex of the cone occurs

Softening temperature (ST): temperature at which the cone has fused down to a spherical lump in which the height is equal to the width of the base

Hemispherical temperature (HT): temperature at which the cone has fused down to a hemispherical lump at which point the height is one-half the width of the base

Fluid temperature (FT): temperature at which the fused mass has spread out in a nearly flat layer with a maximum height of $\frac{1}{16}$ in.

In determining the initial deformation temperature, shrinkage or warping of the cone is ignored if the tip remains sharp. Figure 5.6 illustrates the appearance of the cone before heating and at the above temperatures.

Determination of the Fusibility of Coal Ash. The standard test method for the fusibility of coal and coke ash is ASTM method D 1857. To obtain the ash for the test, coal passing a number 60 (250 μm) sieve (analysis sample prepared in accordance with method D 2013) is heated gradually to a temperature of 800–900°C to remove most of the combustible material. The ash is

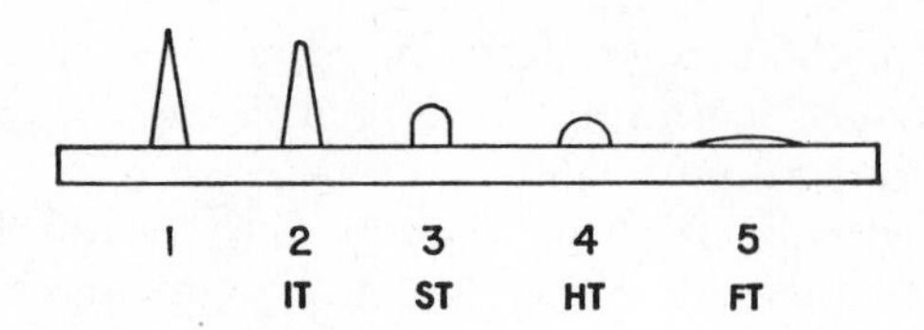

FIGURE 5.6 Critical temperature points. (*Source:* Reprinted with permission from *Annual Book of ASTM Standards,* vol. 05.05, copyright ASTM, 1916 Race St., Philadelphia, PA 19103.)

ground in an agate mortar to pass a number 200 (75μm) sieve, spread on a suitable dish, and ignited in a stream of oxygen for $1\frac{1}{2}$ hours at 800–850° C. Enough coal is used to produce 3–5 g of ash. The ash is mixed thoroughly and moistened with a few drops of dextrin binder and worked into a stiff plastic mass. The mass is then formed into a cone using a cone mold such as that illustrated in Figure 5.7. The cones are dried, mounted on a refractory base, and heated at a specified rate in a gas-fired or electrically heated furnace under either oxidizing or reducing conditions.

In gas-fired furnaces, the atmosphere is controlled by regulating the ratio of air to combustible gas. For reducing conditions an excess of gas over air is maintained, and for oxidizing conditions an excess of air over gas is maintained. Hydrogen, hydrocarbons, and carbon monoxide produce a reducing atmosphere, while oxygen, carbon dioxide, and water vapor are considered to be oxidizing gases. Nitrogen is inert. For a mildly reducing atmosphere, the ratio by volume of reducing gases to oxidizing gases must be maintained between the limits of 20–80 and 80–20 on a nitrogen-free basis. In a gas-fired furnace, this ratio may be difficult to achieve at high temperatures while main-

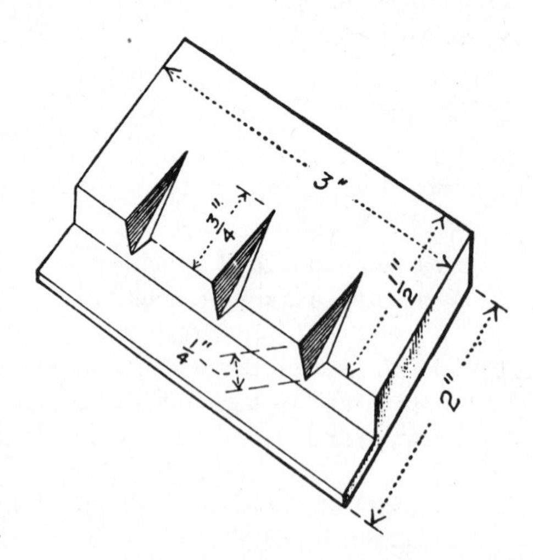

FIGURE 5.7 Brass cone mold. (*Source:* Reprinted with permission from *Annual Book of ASTM Standards,* vol. 05.05, copyright ASTM, 1916 Race St., Philadelphia, PA 19103.)

taining the required temperature rise. For an oxidizing atmosphere, the volume of reducing gases present must not exceed 10%.

In electrically heated furnaces, a mixture of 60 vol % carbon monoxide and 40 vol % ± 5 vol % carbon dioxide produces a reducing atmosphere in the furnace. A regulated stream of air produces an oxidizing atmosphere.

The ASTM method is empirical, and strict observance of the requirements and conditions is necessary to obtain reproducible results. Proper control of the atmosphere surrounding the test specimen is probably the greatest problem encountered in determining ash fusibility, particularly when a reducing atmosphere is used. A mildly reducing atmosphere is specified, since it is believed that this more closely approximates conditions existing in fire beds when coal is burned in several types of combustion equipment. Lower softening temperature values are obtained with a mildly reducing atmosphere than in either strongly reducing or oxidizing atmospheres. With a mildly reducing atmosphere the iron in the ash is present predominantly in the ferrous state, whereas in a strong reducing atmosphere some of the iron may be in the metallic state. In an oxidizing atmosphere the iron is in the ferric state. Both ferric and metallic iron increase the refractory quality of the ash, resulting in higher fusion temperatures. Softening temperature values may vary as much as 150–200° C, depending on the atmosphere in which the test is made.

Temperature measurements are made either with an optical pyrometer or a platinum and platinum-rhodium thermocouple with a high-resistance millivoltmeter. The millivoltmeter or potentiometer should be accurate and readable to 5.5° C (10° F) over the range of 1,000–1,600° C. The temperature-measuring equipment must be properly calibrated by a reliable means. At least once during each week of operation the temperature-measuring equipment should be checked for accuracy by observation of the behavior of small pieces of gold and/or nickel wire with known melting points under routine test conditions. Pure gold (melting point 1,063° C) can be used to calibrate the temperature-measuring equipment in both an oxidizing and a reducing atmosphere. Pure nickel (melting point 1,452° C) can be used only in a reducing atmosphere because it is susceptible to oxidation. This property can be used to advantage in determining whether a furnace that has been set up to operate with a reducing atmosphere is performing properly. Provided the temperature-measuring equipment is calibrated properly, an erratic reading for the melting point of nickel would indicate something other than a reducing atmosphere in the furnace.

In preparing ash for the fusibility test, it is important that the coal be spread out in a thin layer and that adequate circulation of air be maintained during burning. All iron must be converted to the ferric state, and all combustible matter must be removed. A low initial heating temperature and a slow heating rate tend to minimize the retention of sulfur as sulfates in the ash. Following the burning in air, pulverizing the ash and burning it in oxygen will ensure complete conversion of iron to the ferric state and that all combustible material is burned.

Interpretation and Uses of Ash Fusibility Data. Ash fusibility values are often specified in coal contracts because they are believed to be a measure of the tendency of coal ash to form clinkers. Softening temperatures probably are used most often for this purpose. For instance, if it is desirable to have the ash fuse into a large clinker that could be easily removed, then coal with a softening temperature low enough to allow the ash to fuse would be chosen. However, the ash should not soften at a temperature too low, as it may become fluid enough to run through the fire bed and solidify below it, making the ash harder to remove. Coal with high softening temperatures produces ash with relatively small particle size rather than fused masses. Initial deformation and fluid temperatures may also be useful, depending on the type of combustion equipment to be used for burning coal and the manner in which the ash is to be removed.

In practice, types of burning equipment, rate of burning, temperature and thickness of the fire bed, distribution of ash-forming minerals in the coal, and viscosity of the molten ash may influence ash behavior more than do the laboratory-determined ash fusibility characteristics. The correlation of the laboratory test with the actual utilization of coal is only approximate, due to the relative homogeneity of the laboratory test sample compared to the heterogeneous mixture of ash that occurs when coal is burned. Conditions that exist during the combustion of coal are so complex that they are impossible to duplicate completely in a small-scale laboratory test. Therefore, the test should be considered only as an empirical one, and the data should be considered qualitative and should not be overinterpreted.

5.6.6 Free-Swelling Index of Coal

The free-swelling index (FSI) is a measure of the increase in volume of coal when heated under specified conditions. The results from a test may also be used as an indication of the caking characteristics of the coal when it is burned as a fuel. The volume increase can be associated with the plastic properties of coal; coals that do not exhibit any plastic properties when heated do not show free swelling. It is believed that gas formed by thermal decomposition while the coal is in a plastic or semifluid condition is responsible for the swelling. The amount of swelling depends on the fluidity of the plastic coal, the thickness of bubble walls formed by the gas, and interfacial tension between the fluid and solid particles in the coal. When these factors cause more gas to be trapped, greater swelling of the coal occurs.

The free-swelling index of bituminous coals generally increases with an increase in rank, as is illustrated by the data given in Table 5.6 (19). Values for individual coals within a rank may vary considerably. The values for the lower-rank coals are normally less than those for bituminous coals, while anthracite does not fuse and shows no swelling value.

TABLE 5.6 Average Free-Swelling Index Values for Illinois and Eastern Bituminous Coals

Increasing Rank	Coals	Free-Swelling Index
High-volatile C	Illinois Number 6	3.5
High-volatile B	Illinois Number 6	4.5
High-volatile B	Illinois Number 5	5.0
High-volatile A	Illinois Number 5	5.5
High-volatile A	Eastern	6.0–7.5
Medium-volatile	Eastern	8.5
Low-volatile	Eastern	8.5–9.0

Source: Reference 19; reprinted with permission from O.W. Rees, *Chemistry, Uses, and Limitations of Coal Analysis,* Illinois State Geological Survey Report of Investigations no. 220, 1966, pp. 48–49.

Determination of the Free-Swelling Index of Coal. The detailed procedures for determining the free-swelling index of coal are found in ASTM method D 720. In this method, 1 g of the analysis sample is placed in a translucent silica crucible with a prescribed size and shape, and the sample is leveled in the crucible by light tapping on a hard surface. The cold crucible is then lowered into a special furnace and heated to 800° C ± 10° C in $1\frac{1}{2}$ min and 820° C ± 5° C in $2\frac{1}{2}$ min. The test can be made with either gas or electric heating. The button formed in the crucible is then compared to a chart of standard profiles and corresponding swelling index numbers as shown in Figure 5.8. Three to five buttons are made for each sample, and the average of the profile numbers is taken as the free-swelling index.

Some problems associated with the method are the proper heating rate, oxidation or weathering of the coal sample, and an excess of fine coal in the analysis sample. Failure to achieve the proper temperature in the furnace or, more important, the proper heating rate for the sample in the crucible leads to unreliable results. Uneven heat distribution along the walls of the crucible may also cause erratic results. Careful standardization of the equipment used is therefore essential.

Oxidation or weathering of the coal sample leads to a low free-swelling index. To minimize oxidation and the effects on the free-swelling index, samples should be tested as soon as possible after they are collected and prepared. If oxidation of the coal is suspected, the test should be repeated on a known fresh sample of the same coal.

The size consist of the analysis sample may influence the free-swelling index values of some coals. There is evidence that, for many coals, an excess of fine coal (−100−−200 mesh) may cause FSI values to be as much as two index numbers high (19). The amount of fine coal in the analysis sample should be kept at a minimum for this test (and others). Reducing the coal from a large

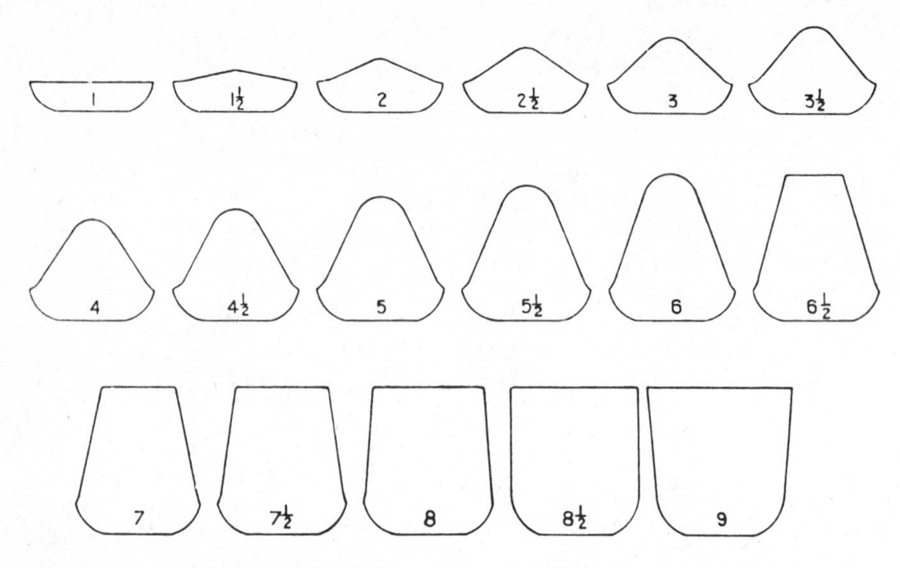

FIGURE 5.8 Profiles and corresponding swelling index numbers not presented at full scale. (*Source:* Reprinted with permission from *Annual Book of ASTM Standards,* vol. 05.05, copyright ASTM, 1916 Race St., Philadelphia, PA 19103.)

particle size to a small particle size in one step tends to produce a high concentration of fine coal. The reduction of coal samples should be done in an appropriate number of steps to avoid this.

Interpretation and Uses of Free-Swelling Index Values. The test for the free-swelling index is an empirical one, and FSI values can be used to indicate the coking characteristics of coal when burned as a fuel. However, these values are not reliable enough for use as parameters in a classification system. Free-swelling index values have been considered useful as an indication of the tendency of coals to form objectionable "coke trees" when burned in certain types of equipment, particularly equipment with underfeed stokers. The decline in the use of underfeed stokers in coal burning equipment along with adjustments of combustion conditions have minimized the problems due to coke tree formation. The use of FSI test data for help in solving this problem has also declined.

Free-swelling index values can be used as an indication of the extent of oxidation or weathering of coals. However, these are not as sensitive to weathering as calorific values.

5.6.7 Plastic Properties of Coal by the Gieseler Plastometer

Testing with a Gieseler plastometer gives a semiquantitative measurement of the plastic property, or apparent melting of coal when heated under prescribed conditions in the absence of air. The chemical nature of the constituents that account for a coal's plastic properties is not known. The material thought to be responsible for the plastic properties of coal has been successfully removed from coal by solvent extraction, leaving a nonplastic residue (20). Such residue has been rendered plastic by returning to it the extracts obtained by the solvent extraction. No definite relationship has been established between the amount of extract and the plastic properties of the coal.

The plastic properties of coal are of practical importance in the coking industry. The test for plasticity is therefore useful in studying coals and blends used in carbonization.

Determination of the Plastic Properties of Coal. The Gieseler plastometer is the one most often used to measure the plastic properties of coal. It consists of a sample holder, a stirrer with four small rabble arms attached at its lower end, a means of applying a known torque to the stirrer, a means of measuring the rate of turning of the stirrer, and a way to heat the sample including provisions for controlling the temperature and rate of temperture rise. The Gieseler retort and furnace assembly is illustrated in Figure 5.9, and a schematic of the retort assembly is shown in Figure 5.10.

The procedure for measuring the plastic properties of coal is found in ASTM method D 2639. In this procedure, 5.0 g of coal passing a number 40(425μm) sieve is placed in the sample holder with the stirrer in place. The coal is packed in and around the stirrer by rotating the stirrer to fill any voids and then applying a weight (10 kg) for 15 min to compress coal around the stirrer. The apparatus is then assembled and immersed in the heating bath, and a known torque is applied constantly and automatically to the stirrer by a magnetic brake system. No movement of the stirrer occurs at first, but as the heating continues, the stirrer begins to turn and its speed increases as the temperature rises. The movement of the stirrer is measured by a magnetic counter actuated by an electric eye or other suitable method. The temperature and rate of movement of the stirrer are observed and recorded throughout the test. The measured values are reported in dial divisions per minute. The values normally determined with the Gieseler plastometer are the following:

1. *Initial softening temperature:* temperature at which the dial movement reaches 1.0 dial division per minute (100 dial divisions = one complete revolution of the stirrer); may be characterized by other rates, but if so, rate must be reported
2. *Maximum fluid temperature:* temperature at which the dial pointer movement (stirrer revolutions) reaches the maximum rate
3. *Solidification temperature:* temperature at which the dial pointer movement stops
4. *Maximum fluidity:* maximum rate of dial pointer movement in dial divisions per minute

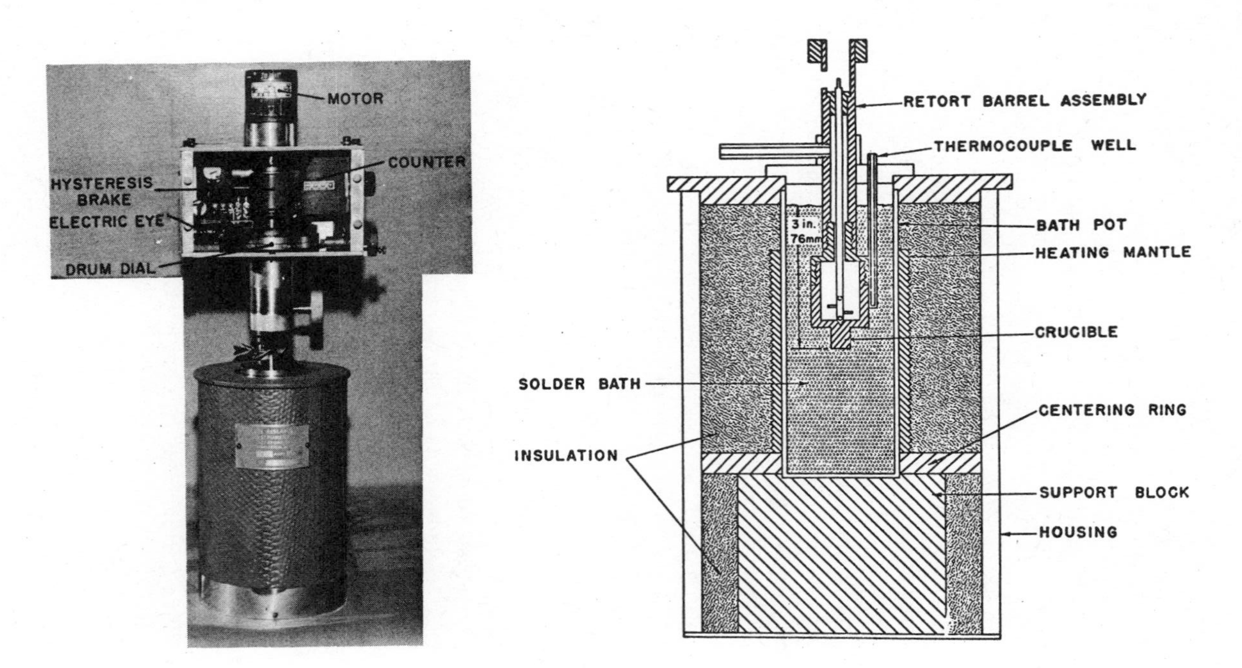

FIGURE 5.9 Gieseler retort and furnace assembly. (*Source:* Reprinted with permission from *Annual Book of ASTM Standards,* vol. 05.05, copyright ASTM, 1916 Race St., Philadelphia, PA 19103.)

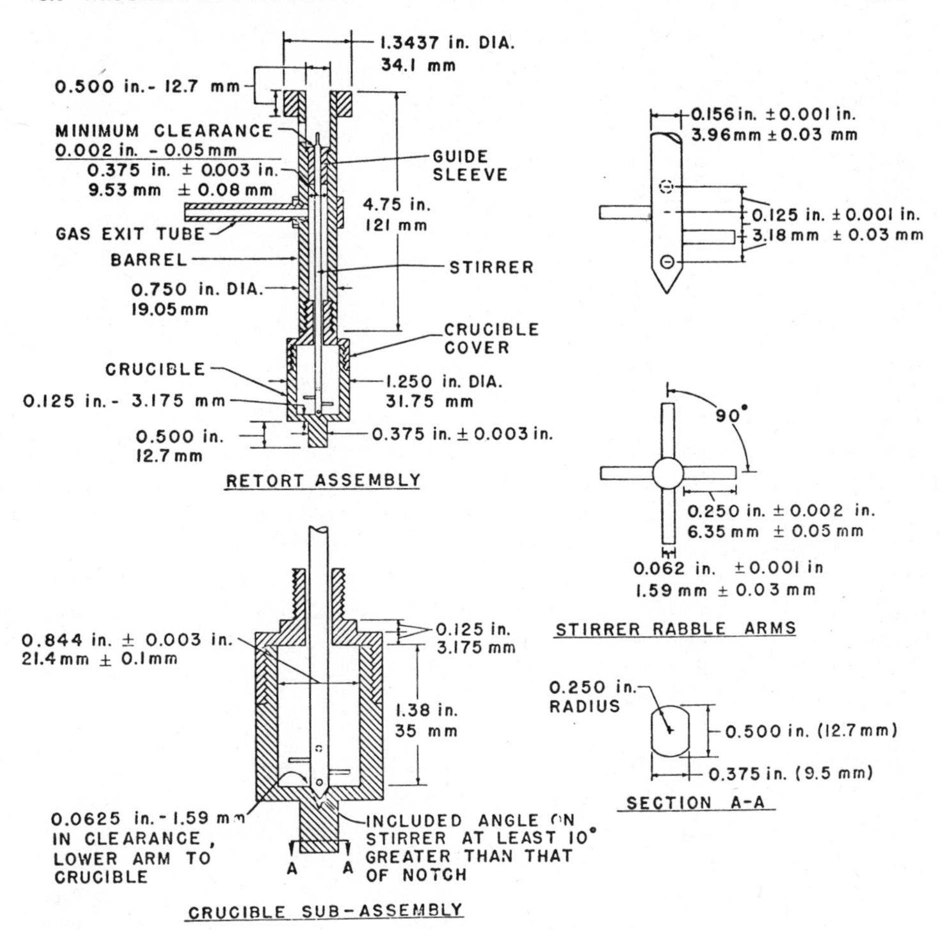

FIGURE 5.10 Retort assembly. (*Source:* Reprinted with permission from *Annual Book of ASTM Standards,* vol. 05.05, copyright ASTM, Philadelphia, PA 19103.)

The sample used in this test should be prepared by air drying a laboratory sample (4 kg crushed to pass a number 4 [4.75 mm] sieve) and then crushed in stages to pass a number 40 (425 μm) sieve. The test should be run as soon as possible after preparing the sample.

The method of measuring the plastic properties of coal is empirical, and strict observance of the requirements and conditions is necessary to obtain repeatable and reproducible values. Many problems are associated with the method, some of which are due to the nature of the coal itself and others to the equipment used. Some problems that arise due to the nature of the coal used are oxidation, packing in the sample holder, and swelling.

Plastic properties are sensitive to the oxidation or weathering of coals. Maximum fluidity is lessened, and extensive oxidation may completely destroy the fluidity of coal. Samples should be tested as soon as possible after they are collected and should be stored under water or in a nonoxidizing atmosphere if there is to be a delay before they are tested.

Proper packing around the stirrer in the plastometer is an important step in the measurement of plastic properties. Some coals may not pack easily due to their weathered condition or to the size consist of the sample. An excess of very fine coal makes the test sample hard to pack.

Some coals swell considerably when heated and may extrude from the sample cup into the barrel of the plastometer. This exerts extra resistance to the stirrer. A well-fitted washer on top of the coal sample may help control the swelling.

Some problems associated with the equipment used are the rate of heating, surface area of the rabble arms on the stirrer, and the manner in which torque is applied. The standard rate of heating influences values obtained in the test, with maximum fluidity being influenced the most. Heating rates higher than the standard lead to higher fluidity values, and lower rates of heating produce lower fluidity values. The plastometer must be thoroughly cleaned after each test. Frequent use and cleaning wear away the stirrer and the rabble arms, gradually decreasing their surface area. As a result, high maximum fluidity values will be obtained. When new, the rabble arms have a total surface area of 136 mm^2. When the surface area decreases to 116 mm^2 (usually after 30–40 tests), the rabble arms should be discarded.

Interpretation and Uses of Plasticity Data. The primary use of plastic property data is for assessing the coking properties of coals. Maximum fluidity values are most often used in this respect, but the plastic range of coals has also been used as a guide for blending coals for carbonization. The plastic range is the temperature between the softening and the solidification temperature. Plastic property data should not be interpreted too closely. These data are probably more useful when applied to low-fluid, less strongly coking coals than in assessing differences in the coking characteristics of high-fluid, more strongly coking coals.

5.6.8 Hardgrove Grindability Index of Coal

Grindability is an index of the relative ease with which a coal may be pulverized in comparison with coals chosen as standards. The Hardgrove method has been accepted as the standard, and ASTM method D 409 is the standard method of grindability of coal by the Hardgrove machine.

Each Hardgrove machine is calibrated by use of standard reference samples of coal with grindability indexes of approximately 40, 60, 80 and 110. These numbers are based on an original soft coal chosen as a standard coal whose grindability index was set at 100. The harder the coal, the lower the index number.

Determination of the Grindability of Coal by the Hardgrove Machine Method. In the Hardgrove machine method, a prepared sample receives a definite amount of grinding energy in a miniature pulverizer, and the change in size consist is determined by sieving. Equipment and materials needed to carry out the test include a Hardgrove grindability machine (Fig. 5.11), standard sieves (with 16.0-, 4.75-, 1.18-, 0.60-, and 0.075-mm openings), a mechanical sieve shaker, and standard reference samples for calibrating the grindability machine (indexes of approximately 40, 60, 80, and 110).

Before the test is run, a sample of coal is collected in accordance with ASTM method D 2234 and prepared according to ASTM method D 2013, except that the sample is not reduced beyond number 4 (4.75-mm) sieve size. A 1,000-g portion of this coal is air dried for 12–48 h and stage crushed to pass a number 16 sieve, with the production of a minimum of material passing a number 30 sieve. A 50 g $\pm$ 0.01 g portion of the 16-by-30 material is evenly distributed in the grinding bowl of the Hardgrove grindability machine. The bowl is fastened into position, and the load fully applied to the driving spindle.

The machine is turned on and allowed to make 60 $\pm$ 0.25 revolutions. The grinding bowl is removed from the machine, and all the coal particles are brushed onto a number 200 sieve with a close-fitting receiving pan. The sieve is covered and shaken mechanically for exactly 10 min, and the underside of the sieve is brushed carefully into the receiver. This shaking and cleaning is repeated for two 5-min periods. The two portions of coal, that remaining on the sieve and that passing the sieve, are weighed separately to the nearest 0.01 g. The grindability index is then determined using a calibration chart obtained by processing the four standard samples. The calibration chart is constructed by plotting, on linear scale coordinates, the calculated weight of material passing a number 200 seive (50 g $\pm$ 0.01 g minus the weight remaining on the number 200 sieve) versus the Hardgrove grindability index of the standard samples.

Failure to obtain duplicate results that fall within the tolerance levels allowed for inter- and intralaboratory comparisons may be due to several factors. The sample moisture may not have been in equilibrium with the laboratory atmosphere; the sample may have been over– or under–air-dried; excessive dust loss may have occurred during screening due to a loose-fitting pan and cover on the sieve; or the sample may not have had an even distribution of particles. The sample should be crushed with a plate mill to obtain an optimum distribution of particles that will pass a number 16 sieve but not a number 30 sieve.

Interpretation and Uses of Hardgrove Grindability Data. The results of grindability measurements with the Hardgrove machine are affected by several factors, including ash content, moisture content, and temperature. Moisture content is particularly troublesome in low-rank coals. The grindability index of coal varies from seam to seam and within the same seam. Grindability data

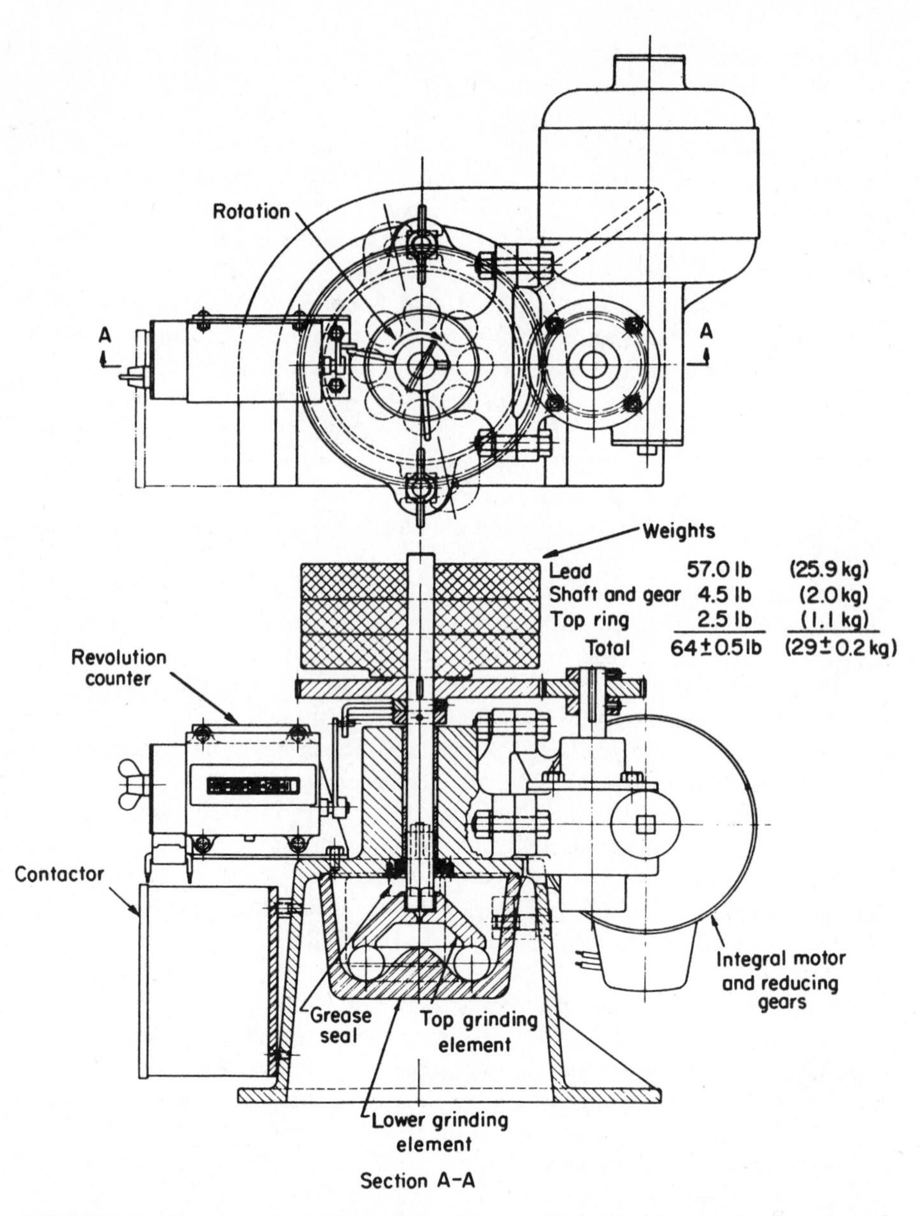

FIGURE 5.11 Hardgrove grindability machine. (*Source:* Reprinted with permission from *Annual Book of ASTM Standards,* vol. 05.05, copyright ASTM, 1916 Race St., Philadelphia, PA 19103.)

are of utmost economic importance to the users of commercial grinding equipment and are used to predict how well coal can be ground for use in various types of combustion equipment.

5.6.9 Composition of Coal Ash

When coal is burned, the mineral constituents form an ash residue composed chiefly of compounds of silicon, aluminum, iron, and calcium, with smaller quantities of compounds of magnesium, titanium, sodium, and potassium. Although the constituents are reported as oxides, they occur in the ash primarily as a mixture of silicates, oxides, and sulfates. The silicates originate in the shale and clay minerals. The principal source of iron oxide is pyrite, which burns to form ferric oxide and sulfur oxides. Calcium and magnesium oxides result from decomposition of carbonate minerals, while the sulfates are formed from interaction between carbonates, pyrite, and oxygen. Examples of the minerals found in coals were given in Table 2.9. Typical limits of the ash composition of bituminous coals (reported as oxides) were given in Table 5.3.

Determination of the Composition of Coal Ash. The following ASTM standard methods pertain to the determination of the composition of coal ash:

Method D 2795: analysis of coal and coke ash

Method D 3682: major and minor elements in coal and coke ash by atomic absorption

Method D 3683: trace elements in coal and coke ash by atomic absorption

Method D 4326: major and minor elements in coal and coke ash by x-ray fluorescence

In the preparation of ash samples for analysis, the slow burning of the coal samples is necessary to prevent the retention of sulfur as sulfate in the ash. If the rate of burning is too rapid, some of the sulfur oxides produced from burning pyrite may react with metal oxides to form stable sulfates. The result is that indefinite amounts of sulfur are retained, which introduces an error into all the analytical results unless all other items are corrected to the SO_3-free basis. In methods D 2795, D 3682, and D 4326, a 3–5 g sample of ash is prepared by placing a coal sample (ground to pass a number 60 sieve) in a cold muffle furnace and gradually heating it so that the temperature reaches 500° C in 1 h and 750° C in 2 h. The mixture is then ground (to pass a number 100 sieve for method D 2795 and a number 200 sieve for methods D 3682 and D 4326) and reignited at 750° C for 1 h. For method D 3683, a coal sample (ground to pass a number 100 sieve) is gradually heated to reach a temperature of 300° C in 1 h and heated at 500° C for an additional hour.

The analysis of coal ash by method D 2795 involves a combination of methods to determine the amounts of SiO_2, Al_2O_3, Fe_2O_3, TiO_2, P_2O_5, CaO, MgO, Na_2O, and K_2O. Two solutions are prepared from the ash. Solution A is obtained by fusing the ash with NaOH followed by a final dissolution of the melt in dilute HCl. This solution is used for the analysis of SiO_2 and Al_2O_3. Solution B is prepared by decomposition of the ash with H_2SO_4, HF, and HNO_3. This solution is used to determine the remaining constituents in the foregoing list. Spectrophotometric procedures are used for the determination of SiO_2, Al_2O_3, Fe_2O_3, TiO_2, and P_2O_5. A chelatometric titration with EDTA is used to determine CaO and MgO, while Na_2O and K_2O are determined by flame photometry.

In ASTM method D 3682, atomic absorption methods are used for the determination of SiO_2, Al_2O_3, Fe_2O_3, CaO, MgO, Na_2O, K_2O, MnO_2, and TiO_2. A spectrophotometric method is used for the determination of P_2O_5. In preparing the ash for analysis, the ash is fused with lithium tetraborate ($Li_2B_4O_7$), and the melt is dissolved in dilute HCl. Solutions containing amounts of the elements suitable for analysis are then prepared from the HCl solution. The atomic absorption spectrophotometer used should be a dual-beam instrument with a deuterium arc background corrector or other comparable simultaneous background correction system. A nitrous oxide–acetylene flame is used for the analysis of SiO_2, Al_2O_3, and TiO_2, whereas an air-acetylene flame is used for Fe_2O_3, CaO, MgO, Na_2O, K_2O, and MnO_2.

In the analysis of major and minor elements in coal and coke ash by x-ray fluorescence, the prepared ash is fused with lithium tetraborate or other suitable flux material. It is then either ground and pressed into a pellet or cast into a glass disk. The pellet or disk is then irradiated with a high energy x-ray beam. The x-radiation that is emitted or fluoresces from the sample is characteristic of the elements in the sample. The x-radiation from the sample is dispersed, and the intensities are measured at selected wavelengths. These intensities are related to the concentrations of the metallic elements in the prepared ash sample as determined by comparison to calibration curves for standard reference materials analyzed under the same conditions. The elements commonly determined by this method include iron, calcium, potassium, aluminum, silicon, phosphorus, magnesium, titanium, and sodium.

In the determination of trace elements in coal or coke ash by atomic absorption according to method D 3683, the coal is ashed and the ash dissolved in aqua regia and HF. A boric acid solution (H_3BO_3) is added to aid in the dissolution. The elements in the solution are then determined by conventional atomic absorption procedures using a dual-beam instrument with a deuterium arc background corrector, or other comparable simultaneous background correction system. Nitrous oxide–acetylene flames are used for the determination of Be, V, and Cr, while air-acetylene flames are used for the determination of Cd, Cu, Mn, Ni, Pb, and Zn.

When coal ash is prepared according to this method (D 3683), the eight elements listed above are quantitatively retained in the ash and are representative of concentrations in the whole coal.

Interpretation and Uses of Ash Composition Data. A compositional analysis of the ash in coal is often useful in the total description of the quality of the coal. Knowledge of the composition of ash is useful in predicting the behavior of ashes and slags in combustion chambers. The amount and composition of ash is important in determining the best cleaning methods for coals, in selecting coals to be used in the production of coke, and in selecting pulverizing equipment to be used in commercial pulverizing operations. Utilization of the ash by-products of coal combustion sometimes depends on the chemical composition of the ash.

A wide range of trace elements occurs in coal, primarily as a part of the mineral matter. The release of certain trace elements into the environment as combustion products or in the disposal of ash is a concern for coal-burning facilities. The determination of certain trace elements in coal and coal ash is becoming an increasingly important part of coal analysis.

The chemical composition of laboratory-prepared coal ash may not be exactly representative of the composition of fly ash, power plant ash, or industrial process ash resulting from the commercial burning of coal. The composition of ash does not give an exact representation of the noncombustible material, or mineral matter, occurring in coal, but it is useful for practical applications.

5.6.10 Analysis of Mineral Matter in Coal

Coals are complex mixtures of organic and inorganic species. The term *mineral matter* refers to the inorganic constituents of coal and is considered to be the sum of all elements that are not part of the organic coal substance (containing carbon, hydrogen, nitrogen, oxygen, and sulfur). The mineral matter includes the minerals associated with the coal and, in addition, the chemically bound elements in the organic portion of coal other than C, H, N, O, and S. The mineral matter is the principal source of the elements that make up the ash when the coal is burned in air or oxygen.

There are several sources of mineral matter in coal. *Inherent mineral matter* generally arises from the plant material from which the coal was formed. This type of mineral matter cannot be easily removed by physical methods, since it is intimately associated with the organic fraction of coal. *Extraneous mineral matter* is composed primarily of quartz, clays, inorganic sulfides, carbonates, and sulfates. This type of mineral matter can usually be separated from the coal during cleaning and washing operations. A partial list of the minerals that have been reported as being associated with coals was given in Table 2.9.

Mineral matter content generally represents a significant proportion of a coal's composition. The amount varies from seam to seam, and values up to 32% have been reported for coals mined in North America. A reasonable value for the "average" amount of mineral matter found in North American coals has been estimated to be 15% (21).

In general, mineral matter is considered both undesirable and detrimental in coal utilization. Its presence affects almost every aspect of mining, preparation, transportation, and utilization. Coal preparation is aimed at reducing the quantity of mineral matter, and the efficient use of the methods chosen depends on its concentration and composition. However, no matter how effective the coal preparation technique, there is always a significant amount of residual mineral matter. This residual material is of considerable importance in coal utilization.

The quality of coke is related to its ash and sulfur content, which are both dependent on the mineral composition of the feed coal. It is thought that the inorganic constituents of coking coals may have a marked effect on the yield of carbonization products and the structure, strength, and reactivity of the resulting coke. The presence of inorganic species can be advantageous, since some of them act as catalysts and thus increase the reactivity of the coke.

When coal is burned in a combustion unit, mineral matter undergoes major changes that lead to problems of clinker formation, fly ash, slagging, and boiler tube corrosion. The efficiency of a combustion unit is related to the amount of ash produced, since it is a diluent. Disposal of the ash can result in large capital expenditures. On the positive side, ash has been utilized as a construction material and is a possible source of refractories. The composition of the ash must be known before it can be utilized in this way.

The increased interest in coal gasification and liquefaction has produced a need for a better understanding of the behavior of minerals in these processes. The possible poisoning of catalysts and the removal of insolubles such as minerals, unreacted coal, char, and insoluble products from the liquefaction product stream are major problems that are encountered. Due to their mineral matter content, the use of certain coals in liquefaction streams and in coal slurries also leads to greater abrasion of valves and pumps.

Determination of the Mineral Matter Content of Coal. The mineral matter content of coal cannot be determined qualitatively or quantitatively from the ash that is formed when the coal is oxidized. The high temperature ashing of coal at 750° C, as designated by ASTM methods, causes a series of reactions involving the minerals in coal. For example, pyrite is oxidized to ferric oxide and sulfur dioxide, carbonates form oxides, and clays lose all water. Quartz is about the only mineral that remains unaltered.

A reliable method of measuring the mineral matter content of a coal is an acid demineralization procedure (22,23). The method depends on the loss of weight of a sample when treated with 40% HF at 50–60° C. Treatment of the sample with HCl before and after the HF treatment helps prevent the retention of insoluble CaF_2 in the coal. Pyrite is not dissolved in the treatment. Consequently, this compound, along with a small amount of residual ash and a small amount of retained chloride (as HCl), must be determined separately. Since two-thirds of the mass of the pyrite (FeS_2) is accounted for by the presence of Fe_2O_3 in the residual ash, the mineral matter content is then given by the formula:

$$MM = \text{weight loss} + HCl + \tfrac{1}{3}(FeS_2) + \text{residual ash} \tag{15}$$

This method has been used with coals of all ranks and requires no assumptions about the nature of the mineral matter. However, it is slow and tedious, and gives only data on the total amount of mineral matter and not its composition.

For analytical purposes, it is desirable to separate the minerals from the coal in an unaltered form. In early studies, density separation methods were used, which were unsatisfactory because of the enrichment of certain minerals in the process. A low-temperature ashing, or plasma ashing, technique has been developed that is more reliable and faster than density separations (24). In this method, low-pressure oxygen is activated by a radio-frequency (rf) discharge. The excited oxygen atoms and other oxygen-containing species oxidize the carbonaceous material at low temperatures (approximately 150° C). The effects of low-temperature ashing and of the oxidizing gas stream on the minerals in coal are minimal. Some pyrite can be oxidized, and to some extent organic sulfur can be fixed as sulfates. The rates of these reactions are functions of operating conditions, such as rf power level and oxygen flow rate.

Factors that affect the rate of low-temperature ashing other than rf power and oxygen flow rate, mentioned above, are the coal particle size and depth of sample bed. Typical conditions for ashing are a particle size of less than 80 mesh, a sample layer density of 30 mg/ cm^2, oxygen flow rate of 100 cm^3/ min, chamber pressure of about 2 torr, and a 50-watt net rf power. The total time required is 36–72 h, and specified conditions must be met during the procedure in order to obtain reproducible results.

Elemental Analysis of the Mineral Matter in Coal. Once the low-temperature ashing procedure has been carried out and the mineral matter residue has been obtained, the minerals can be identified and their concentrations can be determined by a variety of instrumental techniques. Generally, it can be said that no single method yields a complete analysis of the mineral matter in coal. It is often desirable to employ a combination of methods.

One of the most reliable methods used thus far for distinguishing minerals in low-temperature ash (LTA) is x-ray diffraction analysis. However, its application can be limited because of orientation effects, and a reliable method of sample preparation is necessary to prevent these from occurring. X-ray diffraction profiles are determined by using a conventional diffractometer system with monochromatic x-radiation. For qualitative analysis, the specimen is scanned over a wide angular range to ensure all the major diffraction peaks of the component minerals are recorded. Diffraction spacings are then calculated from the peak positions, and the elements present in the sample are determined by using standard tables of diffraction spacings.

X-ray diffraction procedures are used for quantitative analysis of pyrite, calcite, and quartz in low-temperature ashing residues (25,26). Thoroughly mixed calibration mixtures of known proportions of calcite, pyrite, quartz, clay, and an internal standard of CaF_2 are used. Spinning of the sample to remove orientation effects, slow scan rates, and a stabilized x-ray generator and counter are required. The best precision for this type of analysis is only about 10%, due to problems in obtaining uniform mixing of the sample and standards, orientation problems, and difficulty in obtaining representative standards.

Before the development of low-temperature ashing techniques, the infrared analysis of minerals in coals was severely limited because the broad bands of the organic portion of coal overlapped those of the mineral constituents. Since the development of LTA for the removal of the organic fraction of coal, it has been demonstrated that a number of minerals in coal can be identified and analyzed using infrared spectroscopy (27). The spectra are obtained using KBr or CsI pellets containing the finely divided mineral matter. Conventional infrared techniques have been used for successful quantitative measurements of kaolinite and gypsum in mineral matter.

Fourier transform infrared (FTIR) spectroscopy can be successfully applied to the characterization of coals, coal-derived materials, and mineral matter (28,29). It is somewhat limited for identification purposes but can be used for quantitative analysis.

Scanning electron microscopy with an energy-dispersive x-ray system accessory has been used to identify the composition and nature of minerals in coals and to determine the associations of minerals with each other. Examinations can be made on samples resulting from low-temperature ashing techniques or whole coal. With this technique it is possible to identify the elemental components and deduce the mineral types present in coal samples. Computerized systems to evaluate scanning electron microscopy images have been developed and are useful in characterizing the minerals in coal mine dusts, coals, and coal liquefaction residues (26,30). In this system, mineral grains are located, and their elemental compositions are determined by monitoring seven x-ray channels (Al, Si, S, Ca, Fe, K, and Ti). From the various combinations of these elements, it is possible to characterize most of the commonly occurring minerals in coals. Quantitative measurements can be obtained only if the mineral grains are larger than about 1 μm in diameter, due to the limited resolution of the x-ray system.

Optical microscopy is another method that has been used to determine the distribution of minerals in coal. This method is based on the detailed microscopic examination of polished or thin sections of coal in transmitted and/or reflected light. In principal, identification of a mineral type is made by observing several of its optical properties, such as morphology, reflectance, refractive index, and anisotropy. These methods have been widely used by petrographers.

Several methods have been used to determine the trace elements in the mineral matter of coal, as well as in whole coal and coal-derived materials. These methods include spark-source mass spectrometry, x-ray fluorescence, neutron activation analysis, optical emission spectroscopy, and atomic absorption spectroscopy.

Spark-source mass spectrometry (SSMS) has been used extensively in the determination of trace elements in coal. Whole coal samples as well as ash residues, fly ash, respirable coal dust, and lung tissue have been analyzed using this technique. A. G. Sharkey, T. Kessler, and R. A. Friedel (31) analyzed trace elements in 13 coals from 10 coal seams located in Pennsylvania, West Virginia, Virginia, Colorado, and Utah using SSMS. Sixty-four elements ranging in concentration from 0.01–41,000 ppm were determined. Figure 5.12 gives the occurrence frequency of the 64 elements analyzed, and Figure 5.13 gives the concentration ranges of the elements. R. J. Guidoboni (32) compared results from SSMS determination of Mn, Ni, Cr, V, Cu, and Zn with atomic absorption values for the same samples. The relative standard deviations ranged from 6–15% for SSMS and 2–3% for atomic absorption.

X-ray fluorescence analysis is a rapid, simple, and reasonably accurate method of determining the concentration of many minor and trace elements in whole coal. The method is dependent on the availability of suitable standards.

H ND																
Li 100	Be 100											B 100	C ND	N ND	O ND	F 100
Na 100	Mg 100											Al 100	Si 100	P 100	S 100	Cl 100
K 100	Ca 100	Sc 100	Ti 100	V 100	Cr 100	Mn 100	Fe 100	Co 100	Ni 100	Cu 100	Zn 100	Ga 100	Ge 100	As 100	Se 100	Br 100
Rb 100	Sr 100	Y 100	Zr 100	Nb 100	Mo 100	Tc ND	Ru 0	Rh 0	Pd 0	Ag 92	Cd 92	In Standard	Sn 100	Sb 92	Te 85	I 85
Cs 100	Ba 100	La 100	Hf 46	Ta 62	W 69	Re 0	Os 0	Ir 0	Pt 0	Au 0	Hg 38	Tl 31	Pb 100	Bi 31	Po ND	At ND
Fr ND	Ra ND	Ac ND	Ce 100	Pr 100	Nd 100	Pm ND	Sm 100	Eu 100	Gd 85	Tb 85	Dy 85	Ho 77	Er 77	Tm 0	Yb 62	Lu 38
			Th 92	Pa ND	U 92											

FIGURE 5.12 Occurrence frequency of elements in 13 raw coals as determined by spark-source mass spectrometry. All quantities are percentages (ND, not determined; 0, checked but not determined). (*Source:* Reprinted with permission from A.G. Sharkey, Jr., et al., "Trace Elements in Coal Dust by Spark-Source Mass Spectrometry," in *Trace Elements in Fuel,* S.P. Babu, ed., Advances in Chemistry Series no. 141, American Chemical Society, Washington, DC, 1975.)

H ND																
Li 4-163	Be 0.4-3											B 1-230	C ND	N ND	O ND	F 1-110
Na 100-1000	Mg 500-3500											Al 3000-23,000	Si 5000-41,000	P 6-310	S 700-10,000	Cl 10-1500
K 300-6500	Ca 800-6100	Sc 3-30	Ti 200-1800	V 2-77	Cr 26-400	Mn 5-240	Fe 1400-12,000	Co 1-90	Ni 3-60	Cu 3-180	Zn 3-80	Ga 0.3-10	Ge 0.03-1	As 1-10	Se 0.04-0.3	Br 1-23
Rb 1-150	Sr 17-1000	Y 3-25	Zr 28-300	Nb 5-41	Mo 1-5	Tc ND	Ru <0.1	Rh <0.1	Pd <0.1	Ag <0.01-3	Cd <0.01-0.7	In Standard	Sn 1-47	Sb <0.1-2	Te <0.1-0.4	I <0.1-4
Cs 0.2-9	Ba 20-1600	La 0.3-29	Hf <0.3-4	Ta <0.1-8	W <0.1-0.4	Re <0.2	Os <0.2	Ir <0.2	Pt <0.3	Au <0.1	Hg <0.3-0.5	Tl <0.1-0.3	Pb 1-36	Bi <0.1-0.2	Po ND	At ND
Fr ND	Ra ND	Ac ND	Ce 1-30	Pr 1-8	Nd 4-36	Pm ND	Sm 1-6	Eu <0.1-0.4	Gd <0.1-3	Tb <0.1-2	Dy <0.1-5	Ho <0.1-0.4	Er <0.1-0.4	Tm <0.1	Yb <0.1-0.5	Lu <0.1-0.3
			Th <0.1-5	Pa ND	U <0.1-1											

FIGURE 5.13 Concentration range of elements in 13 raw coals analyzed by spark-source mass spectrometry. All quantities are in ppm wt (ND, not determined). (*Source:* Reprinted with permission from A.G. Sharkey, Jr., et al., "Trace Elements in Coal Dust by Spark-Source Mass Spectrometry," in *Trace Elements in Fuel,* S.P. Babu, ed., Advances in Chemistry Series no. 141, American Chemical Society, Washington, DC, 1975.)

Although the major elements in coal—carbon, hydrogen, oxygen, and nitrogen—cannot be analyzed by x-ray fluorescence, most other elements at levels greater than a few parts per million (ppm) are readily determined. This may be the best method of analyzing large numbers of coal samples (33).

Neutron activation analysis techniques are frequently used for trace element analyses of coal and coal-related materials. As many as 61 elements have been examined using instrumental neutron activation analysis (INAA) (34). Precision of the method is 25%, based on all elements reported in coal and other sample matrices. Overall accuracy is estimated at 50% (35).

Emission spectroscopic methods of analysis have been used to determine up to 25 elements present in coal ash. G. B. Dreher and J. A. Schleicher (36) developed methods of analyzing 16 trace elements in high-temperature coal ash. Concentration ranges were from less than 0.3–249 ppm, with detection limits of 0.2–5 ppm, for the 16 elements studied.

Atomic absorption spectroscopy (AAS) is the only instrumental method of analysis for trace elements in coal that is presently accepted by the American Society for Testing and Materials. Other methods of analyzing trace elements are compared to AAS. A discussion of the use of AAS to determine major, minor, and trace elements in coal was given in Section 5.6.9.

A disadvantage is using AAS, as well as most of the other instrumental techniques, is the time required for preparation of the sample. J. E. O'Reilly and M. A. Hale developed a procedure for the direct atomic absorption or emission analysis of powdered whole coal slurries, thereby eliminating time-consuming ashing and sample-digestion procedures. Preliminary results indicate the method is applicable to the estimation of minor-level and trace-level constituents (37,38).

Calculation of Mineral Matter Content of Coal. Determination of a good value for the percent of mineral matter content (% MM) is a very important component of coal analysis. If this quantity cannot be determined directly by the acid demineralization or low-temperature ashing procedure discussed previously, or by other suitable methods, then it is possible to calculate a reasonable value for the mineral matter in coal, provided the necessary data are available.

Several formulas have been proposed for calculating mineral matter in coal, but the two most used formulas are the formula of S. W. Parr (39) and that of J. G. King, M. B. Maries, and H. E. Crossley (40). The Parr formula is the one most often used in the United States and requires only ash and sulfur values as determined in routine analysis:

$$\%\ MM = 1.08A + 0.55S \tag{16}$$

where A = percentage of ash and S = percentage of sulfur. The first term in this formula, 1.08A, is a correction for the loss in weight due to the elimination of water in the decomposition of clay minerals at high temperatures. As mentioned in Section 5.3.1, the water of hydration of mineral matter has been estimated to be 8% of the ash value. The second term in the formula is a correction for the loss in weight when pyrite burns to Fe_2O_3. The Parr formula treats all sulfur as pyritic and makes no allowance for the decomposition of carbonates or fixation of sulfur in the ash.

The King, Maries, Crossley formula is a more elaborate formula that allows for a number of effects:

$$\%\ MM = 1.09A + 0.5S_{pyr} + 0.8CO_2 - 1.1SO_{3_{ash}} + SO_{3_{coal}} + 0.5Cl \tag{17}$$

where A = percentage of ash, S_{pyr} = percentage of pyritic sulfur, CO_2 = percentage of mineral carbon dioxide, $SO_{3_{ash}}$ = percentage of SO_3 in ash, $SO_{3_{coal}}$ = the total sulfur appearing as sulfates in coal, and Cl = percentage of chlorine. In this formula, the various numbers represent correction factors for the loss in weight due to the elimination of water in the decomposition of clay minerals (1.09), for the oxidation of pyrite to Fe_2O_3 and SO_2 (0.5), for the loss of CO_2 from mineral carbonates (0.8), and for the fixation of sulfur in the ash (1.1). The addition of the value representing the sulfate content of the coal sample and one-half the chlorine (assuming one-half the chlorine in coal is found in the mineral matter) completes the formula.

The King, Maries, Crossley formula has been revised by the British National Coal Board, and the final formula is as follows (17):

$$\%\ MM = 1.13A + 0.5S_{pyr} + 0.8CO_2 - 2.8S_{ash} + 2.8S_{coal} + 0.5Cl \tag{18}$$

With this formula, a reasonably accurate value of the mineral matter can be calculated, but many parameters need to be determined in order to perform the computation.

Interpretation and Uses of Mineral Matter Data. An ultimate analysis that can claim to represent the composition of the organic substance of a coal is said to be on the *dry, mineral-matter-free* (dmmf) basis. The dmmf basis is a hypothetical condition corresponding to the concept of a pure coal substance. Since the dry, ash-free basis for coal neglects the changes in mineral matter when coal is burned, the dmmf basis is preferred whenever the mineral matter can be determined or calculated.

The ASTM method of classifying coals depends on the calculation of the volatile matter yield and fixed carbon values on the dmmf basis. Calorific values are calculated on the moist, *mineral-matter-free basis.* The Parr formula is used in the ASTM system to calculate the mineral matter from ash and sulfur data.

5.7 COAL CLASSIFICATION

Due to the worldwide occurrence of coal deposits, the numerous varieties of coal that are available, and its many uses, many national coal classification systems have been developed. These systems often are based on characteristics of domestic coals without reference to coals of other countries. Terms for describing similar or identical coals are not used uniformly in these various systems.

Efforts have been made in the United States and abroad to develop systems for classifying coal that are based on characteristic properties determined by laboratory methods. Attempts have also been made to develop an international system for classifying coal to eliminate confusion in international trade and to facilitate the exchange of technical and scientific information related to coal utilization and research.

A discussion of the system used for classifying coals in the United States and the international systems of coal classification follows.

5.7.1 Classification of Coal by Rank

In ASTM standard D 388, "Classification of Coals by Rank," coals are classified according to their degree of metamorphism, or progressive alteration, in the natural series from lignite to anthracite. The basis for the classification is according to fixed carbon and calorific values calculated to the mineral-matter-free basis. Higher-rank coals are classified according to fixed carbon on the dry mineral-matter-free basis. Lower-rank coals are classed according to their calorific values on the moist mineral-matter–free basis. The agglomerating character is also used to differentiate certain classes of coals.

To classify a coal according to this system, the calorific value and a proximate analysis (moisture, ash, volatile matter, and fixed carbon by difference) are needed. For lower-rank coals, the equilibrium moisture must also be determined. To calculate these values to the mineral-matter–free basis, the following Parr formulas (38) are used:

$$\text{Dry, MM-free FC} = \frac{(\text{FC} - 0.15\text{S})}{[100 - (\text{M} + 1.08\text{A} + 0.55\text{S})]} \times 100 \qquad (19)$$

$$\text{Dry, MM-free VM} = 100 - \text{Dry, MM-free FC} \qquad (20)$$

$$\text{Moist, MM-free Btu} = \frac{\text{Btu} - 50\text{S}}{[100 - (1.08 + 0.55\text{S})]} \qquad (21)$$

where MM = percentage of mineral matter, Btu = calorific value, in Btu/lb, FC = percentage of fixed carbon, VM = percentage of volatile matter, M = percentage of moisture, A = percentage of ash, and S = percentage of sulfur. These quantities are all on the inherent- or equilibrium-moisture basis. This basis pertains to coal containing its natural inherent, or bed, moisture but not including any surface moisture. Sampling procedures used are to be those that are most likely to preserve the inherent moisture.

Coals are classified by rank according to the information given in Table 5.7. Coals with fixed carbon values of 69% or more, as calculated on the dry, mineral-matter–free basis, are classified according to their fixed carbon values. Coals with calorific values less than 14,000 Btu/lb, as calculated on the moist, mineral-matter–free basis, are classified according to their calorific values on a moist, mineral-matter–free basis, provided their dmmf fixed carbon is less than 69%. The agglomerating character is considered for coals with 86% or more dmmf fixed carbon and for coals with calorific values between 10,500 and 11,500 Btu/lb, as calculated on the moist, mineral-matter–free basis.

The ASTM system provides for the classification of all ranks of coal, whereas the international classification is based on two systems, one for the hard coals and the other for brown coals and lignites. The borderline between the two systems has been set at 10,260 Btu/lb calculated on a moist, ash-free basis. Hard coals are those with Btu values above 10,260 Btu/lb.

The International Classification of Hard Coals by Type System is based on the dry, ash-free volatile matter; the calorific value expressed on a moist, ash-free basis; and the coking and caking properties. A coal is given a three-figure code number from a combination of these properties. Table 5.8 lists the classification parameters and the development of numerical symbols to represent the groups and subgroups (41).

Coals are first divided into classes 1–5, containing coals with volatile matter

TABLE 5.7 Classification of Coals by Rank

Class	Group	Fixed Carbon Limits (%, dry, mineral-matter–free basis)		Volatile Matter Limits (%, dry, mineral-matter–free basis)		Calorific Value Limits, (Btu/lb, moist, mineral-matter–free basis)[a]		Agglomerating Character
		Equal to or Greater Than	Less Than	Greater Than	Equal to or Less Than	Equal to or Greater Than	Less Than	
I. Anthracitic	1. Meta-anthracite	98	—	—	2	—	—	Nonagglomerating
	2. Anthracite	92	98	2	8	—	—	Nonagglomerating
	3. Semianthracite[b]	86	92	8	14	—	—	Nonagglomerating
II. Bituminous	1. Low-volatile bituminous coal	78	86	14	22	—	—	Commonly agglomerating[d]
	2. Medium-volatile bituminous coal	69	78	22	31	—	—	Commonly agglomerating[d]
	3. High-volatile A bituminous coal	—	69	31	—	14,000[c]	—	Commonly agglomerating[d]
	4. High-volatile B bituminous coal	—	—	—	—	13,000	14,000	Commonly agglomerating[d]
	5. High-volatile C bituminous coal	—	—	—	—	11,500	13,000	Commonly agglomerating[d]
						10,500	11,500	Agglomerating
III. Subbituminous	1. Subbituminous A coal	—	—	—	—	10,500	11,500	Nonagglomerating
	2. Subbituminous B coal	—	—	—	—	9,500	10,500	Nonagglomerating
	3. Subbituminous C coal	—	—	—	—	8,300	9,500	Nonagglomerating
IV. Lignite	1. Lignite A	—	—	—	—	6,300	8,300	Nonagglomerating
	2. Lignite B	—	—	—	—	—	6,300	Nonagglomerating

Source: Reprinted with permission from *Annual Book of ASTM Standards,* vol. 05.05, copyright ASTM, 1916 Race St., Philadelphia, PA 19103.

Note: This classification does not include a few coals, principally nonbanded varieties, that have unusual physical and chemical properties and that come within the limits of the fixed carbon or calorific value of the high-volatile bituminous and subbituminous ranks. All of these coals either contain less than 48% dry, mineral-matter–free fixed carbon or have more than 15,500 moist, mineral-matter–free British thermal units per pound.

[a] Moist refers to coal containing its natural inherent moisture but not including visible water on the surface of the coal.

[b] If agglomerating, classify in low volatile group in the bituminous class.

[c] Coals having 69% or more fixed carbon on the dry, mineral-matter–free basis shall be classified according to fixed carbon, regardless of calorific value.

[d] It is recognized that there may be nonagglomerating varieties in these groups of the bituminous class, and there are notable exceptions in the high-volatile C bituminous group.

(dry, ash-free basis) up to 33%. Coals with volatile matter greater than 33% are contained in classes 6–9, separated according to their gross calorific value on the moist, ash-free basis. Although the moist calorific value is the primary parameter for classes 6–9, the volatile matter does continue to increase with the rising class number.

The classes of coal are subdivided into groups according to their coking properties, as reflected in the behavior of the coals when heated rapidly. A broad correlation exists between the crucible swelling number and the Roga index (International Organization for Standardization [ISO] Methods), and either of these may be used to determine the group number of a coal.

Coals classified by class and group are further subdivided into subgroups, defined by reference to coking properties. The coking properties are determined by either the Gray-King coke-type assay or the Audibert-Arnu dilatometer test (International Organization for Standardization Methods). These tests express the behavior of a coal when heated slowly, as in carbonization.

In the three-figure code number that describes the properties of a coal, the first digit represents the class number, the second is the group number, and the third is the subgroup number. The international classification accommodates a wide range of coals through the use of the nine classes and various groups and subgroups.

Brown coals and lignites have been arbitrarily defined for classification purposes as those coals having a moist, ash-free calorific value less than 10,260 Btu/lb. These are classified by a code number that is a combination of a class number and a group number. The class number represents the total moisture of the coal as mined and the group number the percentage tar yield from the dry, ash-free coal. Table 5.9 illustrates this classification system for brown coals and lignites (41).

5.8 NEW DEVELOPMENTS IN INSTRUMENTS FOR ROUTINE COAL ANALYSIS

The introduction of microprocessors and microcomputers in recent years has led to the development of a new generation of analytical instruments for coal analysis. In particular, automated instrumentation has been introduced that can determine the moisture, ash, volatile matter, carbon, hydrogen, nitrogen, sulfur, oxygen, and ash fusion temperatures in a fraction of the time required to complete most of the ASTM standard procedures.

The microprocessor-controlled instruments have been developed for the multiple-sample proximate analysis of coal. One of these is a semiautomated instrument consisting of an oven and a furnace mounted on a control console with an auxiliary printer. The unit can also be interfaced with an electronic balance (42). The oven is used for determining the moisture, and the furnace is used for determining the volatile matter and ash. The microprocessor is used to save data and perform calculations as well as to control the oven and fur-

TABLE 5.8 International Classification of Hard Coal by Type

Groups (determined by coking properties)			Code Numbers (The first figure of the code number indicates the class of the coal, determined by volatile-matter content up to 33% VM and by calorific parameter above 33% VM. The second figure indicates the group of coal, determined by coking properties. The third figure indicates the subgroup, determined by coking properties.)										Subgroups (determined by coking properties)		
Group Number	Alternative Group Parameters												Subgroup Number	Alternative Subgroup Parameters	
	Free-Swelling, Index (crucible-swelling number)	Roga Index												Dilatometer	Gray-King
3	>4	>45					435	535	635				5	>140	$>G_8$
						334	434	534	634				4	>50–140	G_5–G_8
						333	433	533	633	733			3	>0–50	G_1–G_4
						332 a / 332 b	432	532	632	732	832		2	≤0	E–G
2	$2\frac{1}{2}$–4	>20–45				323	423	523	623	723	823		3	>0–50	G_1–G_4
						322	422	522	622	722	822		2	≤0	E–G

							321	421	521	621	721	821		1	Contraction only	B–D
1	1–2	>5–20				212	312	412	512	612	712	812		2	≤0	E–G
						211	311	411	511	611	711	811		1	Contraction only	B–D
0	0–$\frac{1}{2}$	0–5		100 a	100 b	200	300	400	500	600	700	800	900	0	Nonsoftening	A
Class Number			0	1		2	3	4	5	6	7	8	9	As an indication, the following classes have an approximate volatile-matter content of Class 6: 33–41% volatile matter 7: 33–44% " " 8: 35–50% " " 9: 42–50% " "		
Class Parameters	Volatile Matter (dry, ash-free)		0–3	>3–10 >3–6.5	>3–10 >6.5–10	>10–14	>14–20	>20–28	>28–33	>33	>33	>33	>33			
Class Parameters	Calorific Parameter[a]		—	—		—	—	—	—	>13,950	>12,960–13,950	>10,980–12,960	>10,260–10,980			
Classes (determined by volatile matter up to 33% VM and by calorific parameter above 33% VM)																

Source: Reprinted with permission from H.H. Lowry, ed., *Chemistry of Coal Utilization,* supplementary vol., John Wiley & Sons, Inc., 1963.

Note: (1) Where the ash content of coal is too high to allow classification according to the present systems, it must be reduced by laboratory float-and-sink methods (or any other appropriate means). The specific gravity selected for flotation should allow a maximum yield of coal with 5–10% ash. (2) 332a . . . >14–16 pct volatile matter; 332 b . . . >16–20 pct volatile matter. (3) Classes determined by volatile matter up to 33 pct volatile matter and by calorific parameter above 33 pct volatile matter.

[a]Gross calorific value on moist, ash-free basis (30° C, 96 pct relative humidity, Btu/lb).

TABLE 5.9 International Classification of Coals with a Gross Calorific Value Below 10,260 Btu/lb* (Statistical Grouping)

Group Number	Group Parameter tar yield (dry, ash-free) %	Class Parameter; Total moisture, ash-free, percent					
		Class 10 0–20%	Class 11 20–30%	Class 12 30–40%	Class 13 40–50%	Class 14 50–60%	Class 15 60–70%
40	>25	1040	1140	1240	1340	1440	1540
30	20–25	1030	1130	1230	1330	1430	1530
20	15–20	1020	1120	1220	1320	1420	1520
10	10–15	1010	1110	1210	1310	1410	1510
00	10 and less	1000	1100	1200	1300	1400	1500

Source: Reprinted with permission from H.H. Lowry, ed., *Chemistry of Coal Utilization,* supplementary vol., John Wiley & Sons, Inc., 1963.

Note: The total moisture refers to freshly mined coals.

*Moist, ash-free basis (30°C and 96% relative humidity).

nace times and temperatures. In the analysis of coals with this instrument, the moisture values are determined as the loss in mass after drying the samples in fused silica crucibles for 60 min at 107° C in the moisture oven. Dry, heated air is allowed to flow over the samples during the drying process. Lids are then placed on the dried coal samples, which are then placed in the furnace. The temperature of the furnace is then raised to 950° C at a rate of 35° C/min. After 7 min at this temperature, the samples are removed and cooled before weighing. The volatile matter is the loss in mass during this heating procedure. The crucible lids are removed before the samples are placed in the furnace again for the ash determination. The results of the analysis can then be calculated with the microprocessor and printed. In the determination of volatile matter, the furnace can be preheated to 950° C before inserting the samples into the furnace (43).

The other microprocessor-controlled instrument developed for multiple-sample proximate analysis of coal is a fully automated system (44). The microprocessor controls most of the functions and performs all the calculations in the analysis. The unit has a single oven-furnace with a turntable for the sample crucibles and an internal balance for weighing the samples during the analysis. In the analysis of coals, samples are loaded into the tared crucibles, and the moisture is determined by heating the samples in a nitrogen atomsphere at 107° C until a constant weight is obtained. The complete set of crucibles is weighed every 3 min by the internal balance. After the moisture determination, lids are placed on the fused silica crucibles and the temperature of the furnace is raised to 950° C at a rate of 50° C/min. The samples are held at this temperature for 7 min and then weighed. The volatile matter is the mass lost during this heating step. The temperature of the furnace is dropped to 600° C, the lids are removed from the crucibles, and the samples are ashed by raising the temperature of the furnace to 750° C and heating the samples in oxygen until a constant weight is obtained. The control unit for the instrument can be used to print the results of the analysis in various formats.

The proximate analysis instruments give moisture, ash, and volatile matter values that compare favorably with the values obtained by the ASTM standard methods. These instruments are currently being studied by an ASTM Task Group for the development of a standard method for instrumental proximate analysis.

Several microprocessor-controlled instruments have been developed for the simultaneous determination of carbon, hydrogen, and nitrogen in various samples. An ASTM Task Group is currently examining these instruments for possible use in a standard method for the instrumental determination of carbon, hydrogen, and nitrogen in coal (45). Some basic requirements for the instruments are that they provide for the complete conversion of the carbon, hydrogen, and nitrogen in coal to carbon dioxide, water vapor, and elemental nitrogen, and for the quantitative determination of these gases in an appropriate gas stream. There are essentially three configurations for the instruments available, depending on the detection scheme employed by each instrument.

In one configuration, the combustion gases are conducted through a series of thermal conductivity detectors and gas absorbers aligned in a way that allows for the detection of the appropriate gas. The combustion gases must first be treated to remove halides, sulfur oxides, and residual oxygen from the gas stream, and to reduce nitrogen oxides to nitrogen. At the water vapor detector, the gases pass through the sample side of the detector, a water vapor absorber, and then the reference side of the detector. At the carbon dioxide detector, the gases are conducted through the sample side of the detector, a carbon dioxide absorber, and then the reference side of the detector. Finally, the remaining gas stream, containing only nitrogen and the carrier gas, is passed through the sample side of the detector and vented. The carrier gas serves as the reference in this detector.

In a second configuration, halides and sulfur oxides are removed from the combustion gases before they are collected in a ballast tank. Aliquots of the combustion gases are then taken for analysis. The carbon dioxide and water vapor are determined by infrared detection using precise wavelength windows so that the absorbances are due only to these gases. Nitrogen is determined using a thermal conductivity detector. This is accomplished by taking an aliquot of the combustion gases, mixing it with the carrier gas (helium), and treating the gas stream to reduce nitrogen oxides to nitrogen and to remove oxygen, carbon dioxide, and water vapor. Figure 5.14 is a schematic of this system.

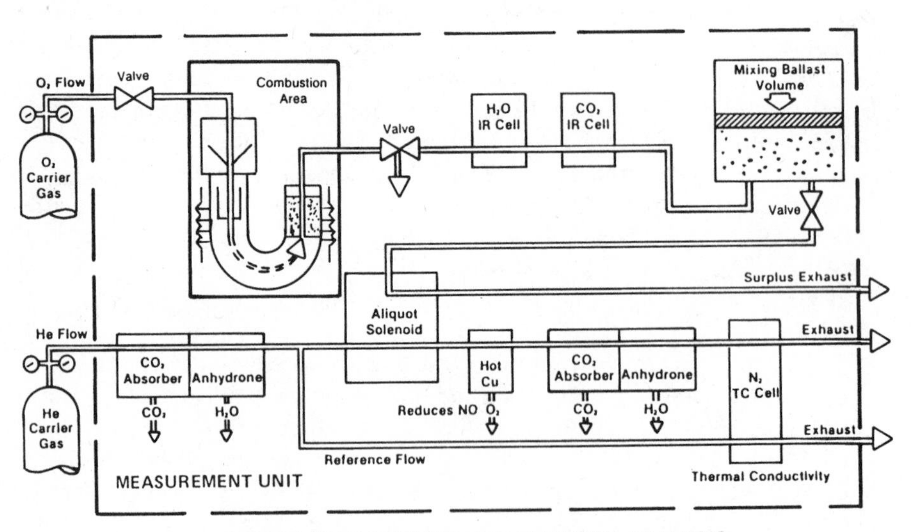

FIGURE 5.14 Apparatus for the determination of carbon and hydrogen by the infrared detection method and nitrogen by the thermal conductivity method. (*Source:* Courtesy of LECO Corporation, 3000 Lakeview Ave., St. Joseph, MI 49085.)

The third configuration is essentially a modified gas chromatographic system. In this system, the combustion gases are treated to reduce all nitrogen oxides to nitrogen and to remove halides and sulfur oxides. The treated gas stream is then passed through a chromatographic column. Nitrogen, carbon dioxide, and water vapor are determined at appropriate retention times by thermal conductivity detection.

A disadvantage of some of the instrumental methods for determining carbon, hydrogen, and nitrogen is the small sample size used in the analysis. A typical sample size for some of the instruments is 1–3 mg, due to the detection systems used and the requirements for scrubbing the gas streams. The previously described system that employs infrared detection for the determination of carbon dioxide and water vapor can accommodate a much larger sample. A 100-mg sample is used with this instrument. The larger size increases the probability that the sample is representative of the quantity of coal being analyzed.

The rapid proximate analysis of coal, coke, and coal-derived products using microcomputer-controlled thermogravimetric analysis (TGA) has been studied by several labs (46–48). In a typical procedure using this method, the low-mass TGA furnace containing the coal sample is heated rapidly to 110°C and held at this temperature for 5 min. A nitrogen atmosphere is maintained inside the furnace. The loss in mass from the coal sample represents the moisture. Following the moisture determination, the furnace is heated rapidly (100°C/min) to 950°C and held at this temperature for 7 min. An atmosphere of nitrogen is maintained in the furnace, and the loss in mass represents the volatile matter lost from the coal. After this determination, the atmosphere is switched from nitrogen to air, and the coal residue is combusted, leaving the ash. The fixed carbon is then calculated by subtracting the values of the moisture, volatile matter, and ash from 100. The total time required for the proximate analysis of one coal sample using TGA is 20–30 min. This method of analysis is currently being studied by an ASTM Task Group as a possible method for the rapid proximate analysis of coal and coke.

New developments in ash fusion furnaces have also been made in recent years. Some furnace systems that have been introduced use a high resolution television camera, a monitor, and a videotape recorder to record the melting of ash cones in a stationary furnace. The operator can then view the videotape to record the ash fusion temperatures. Another ash fusion furnace system, which is more automated, uses a rotating pedestal inside the ash fusion furnace to present the ash cones to the measurement window of a charged coupled device (CCD) once each 6 s. As each cone is scanned, the CCD effectively draws a digital photograph of the cone. Each subsequent scan is compared to the previous scan, and temperatures are automatically recorded when the height of each melting cone reaches a preset level (49). Figure 5.15 is a schematic of this system.

The introduction of the microprocessor and its subsequent incorporation in analytical instrumentation has revolutionized the coal testing and analysis

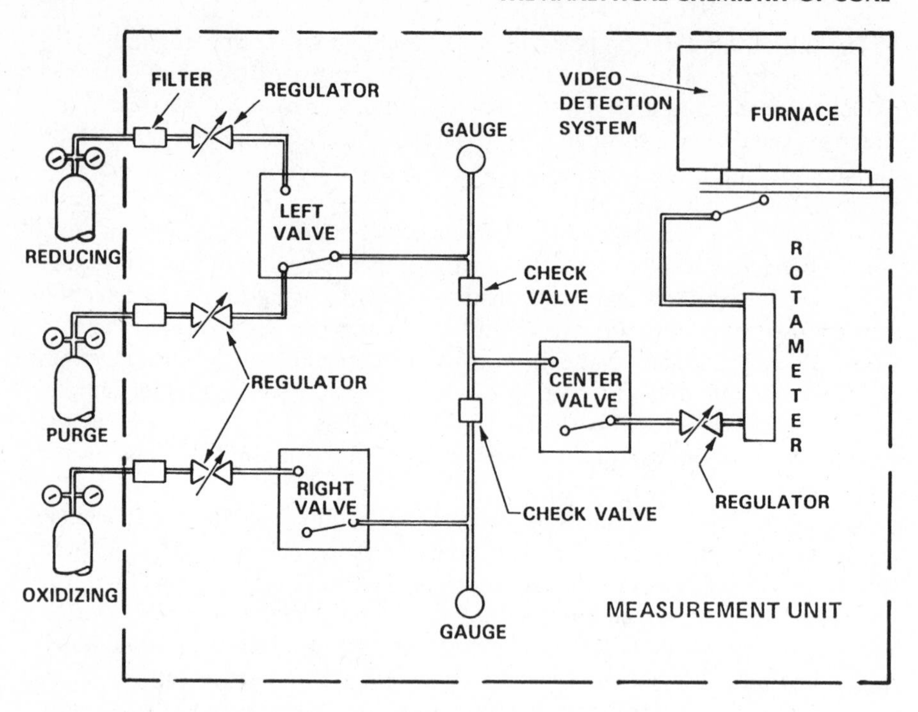

FIGURE 5.15 Ash fusion apparatus with video detection system. (*Source:* Courtesy of LECO Corporation, 3000 Lakeview Ave., St. Joseph, MI 49085.)

industry. Many of the new instruments introduced in recent years are truly automatic and require little operator assistance. If the instruments are maintained in good operating condition and properly calibrated, one can obtain better repeatability of results when using them. Most of the methods used by these instruments are empirical, however, and the accuracy of the results are highly dependent on the quality and suitability of the standards used to standardize the instruments.

SUMMARY

Many problems are associated with the analysis of coal. Its heterogeneous nature, the tendency of coal to gain or lose moisture and to undergo oxidation when exposed to the atmosphere, and the large number of tests and analyses required to adequately characterize a coal are some of the problems.

The American Society for Testing and Materials, with the cooperation of representatives from all areas of the coal industry, has developed standard methods of analysis for coal. Many of these tests are empirical in nature, and strict adherence to the procedural guidelines is necessary in order to obtain repeatable and reproducible results. The type of analysis normally requested in the coal industry may be a proximate analysis (moisture, ash, volatile matter, and fixed carbon) or an ultimate analysis (carbon, hydrogen, sulfur, nitrogen, oxygen, and ash). Quite often a variation of a proximate analysis or an ultimate analysis is requested, along with one or more of the miscellaneous analysis or tests discussed in this chapter.

Restrictions that have been placed on the coal used in coal-fired power plants and other coal-burning facilities have created a need for more coal analyses as well as a need for more accurate and faster methods of analysis. This trend will continue, and more testing will be required with the increased use of coal in liquefaction and gasification plants.

REFERENCES

General References

Annual Book of ASTM Standards, vol. 05.05, American Society for Testing and Materials, Philadelphia.

M.A. Elliott, ed., *Chemistry of Coal Utilization,* 2d supplementary vol., Wiley, New York, 1981.

C.K. Karr, Jr., ed., *Analytical Methods for Coal and Coal Products,* vols. 1, 2, and 3, Academic, New York, 1978–1979.

H.H. Lowry, ed., *Chemistry of Coal Utilization,* supplementary vol., Wiley, New York, 1963.

O.W. Rees, *Chemistry, Uses, and Limitations of Coal Analysis,* Illinois State Geological Survey Report of Investigations no. 220, Urbana, 1966.

Staff, Office of the Director of Coal Research, *Methods of Analyzing and Testing Coal and Coke,* U.S. Bureau of Mines Bulletin no. 638, Pittsburgh, 1967.

Specific References

1. R.F. Abernethy, S. Ergun, R.A. Frieldel, J.T. McCartney, and I. Wender, "Coal and Coke," in F.D. Snell and L.S. Ettre, eds., *Encyclopedia of Industrial Chemical Analysis,* vol. 10, Wiley, New York, 1970, p. 210.
2. *Annual Book of ASTM Standards,* vol. 05.05, American Society for Testing and Materials, Philadelphia.
3. O.W. Rees, *Chemistry, Uses, and Limitations of Coal Analysis,* Illinois State Geological Survey Report of Investigations no. 220, Urbana, 1966, p. 13.
4. W.H. Ode, "Coal Analysis and Mineral Matter," in H.H. Lowry, ed., *Chemistry of Coal Utilization,* supplementary vol., Wiley, New York, 1963, p. 204.
5. Rees, p. 17.
6. W.J. Montgomery, "Standard Laboratory Test Methods for Coal and Coke," in C.K. Karr, Jr., ed., *Analytical Methods for Coal and Coal Products,* vol. 1, Academic, New York, 1978, p. 205.
7. Ode, p. 209.

8. K.J. Thrasher, S.M. Williams, and J.T. Riley, *Proc. Collegiate Assoc for Mining Educ., Conf.*, Lexington, KY, 1984, pp. 153–170.
9. Rees, p. 30.
10. Ode, p. 213.
11. N.F. Shimp, R.J. Helfinstine, and J.K. Kuhn, *Prepr., Div. Fuel Chem., Am. Chem. Soc., 20*(2), 99–107 (1975).
12. D.W. Van Krevelen, *Coal,* Elsevier, Amsterdam, 1961, pp. 161–170.
13. M. Schütze, *Z. Anal. Chem., 118,* 245–258 (1959).
14. J. Unterzaucher, *Analyst, 77,* 584 (1952).
15. I.J. Oita and H.S. Conway, *Anal. Chem., 26,* 600 (1954).
16. Staff, Office of the Director of Coal Research, *Methods of Analyzing and Testing Coal and Coke,* U.S. Bureau of Mines Bulletin no. 638, Pittsburgh, 1967, pp. 19–20.
17. J.N. Chakrabarti, "Methods of Determining Chlorine in Different States of Combination in Coal," in C.K. Karr, Jr., ed., *Analytical Methods for Coal and Coal Products,* vol. 1, Academic, New York, 1978, pp. 325–326.
18. J.K. Kuhn, "The Determination of Forms of Sulfur in Coal and Related Materials," in T.D. Wheelock, ed., *Coal Desulfurization,* ACS Symposium Series no. 64, American Chemical Society, Washington, 1977, pp. 16–21.
19. Rees, pp. 48–49.
20. E.D. Pierron and O.W. Rees, *Solvent Extract and the Plastic Properties of Coal,* Illinois State Geological Survey Circular no. 288, Urbana, 1960, pp. 1–12.
21. H.J. Gluskoter, "Mineral Matter and Trace Elements in Coal," in S.P. Babu, ed., *Trace Elements in Fuel,* Advances in Chemistry Series no. 141, American Chemical Society, Washington, 1975, pp. 1–22.
22. W. Radmacher and P. Mohrhauer, *Brennst, Chem., 36,* 236 (1955).
23. M. Bishop and D.L. Ward, *Fuel, 37,* 191 (1958).
24. H.J. Gluskoter, N.F. Shimp, and R.R. Ruch, "Coal Analysis, Trace Elements, and Mineral Matter," in M.A. Elliott, ed., *Chemistry of Coal Utilization,* 2d supplementary vol., Wiley New York, 1981, p. 411.
25. C.P. Rao and J.J. Gluskoter, *Occurrence and Distribution of Minerals in Illinois Coals,* Illinois State Geological Survey Circular no. 476, Urbana, 1973, 56 p.
26. P.L. Walker, Jr., W. Spackman, P.H. Given, A. Davis, R.G. Jenkins, P.C. Painter, *Characterization of Mineral Matter in Coals and Coal Liquefaction Residues,* Final Report to Electric Power Research Institute, RP 366-1, 779-19, Palo Alto, CA, 1980, pp. 2.24–2.27.
27. P.A. Estepp, J.J. Dovach, and C.K. Karr, Jr., *Anal. Chem., 40,* 358–363 (1968).
28. P.C. Painter, M.M. Coleman, R.G. Jenkins, and P.L. Walker, Jr., *Fuel, 57,* 125–126 (1978).
29. P.C. Painter, M.M. Coleman, R.G. Jenkins, P.W. Whang, and P.L. Walker, Jr., *Fuel, 57,* 337–344 (1978).
30. P.L. Walker, pp. 2.29–2.30.
31. A.G. Sharkey, Jr., T. Kessler, and R.A. Friedel, "Trace Elements in Coal Dust by Spark-Source Mass Spectrometry," in Babu, pp. 48–56.
32. R.J. Guidoboni, *Anal. Chem., 45,* 1275–1277 (1973).
33. J.K. Kuhn, W.F. Harfst, and N.F. Shimp, "X-ray Fluorescence Analysis of Whole Coal," in Babu, pp. 66–73.
34. R.R. Ruch, R.A. Cahill, J.K. Frost, L.R. Camp, and H.J. Gluskoter, *J. Radioanal. Chem., 38,* 415–424 (1977).
35. D.W. Sheibly, "Trace Elements by Instrumental Neutron Activation Analysis for Pollution Monitoring," in Babu, pp. 98–117.

36. G.B. Dreher and J.A. Schleicher, "Trace Elements in Coal by Optical Emission Spectroscopy," in Babu, pp. 35–47.
37. J.E. O'Reilly and M.A. Hale, *Anal. Lett.*, *10*(13), 1095–1104 (1977).
38. J.E. O'Reilly and D.G. Hicks, *Anal. Chem.*, *51*, 1905–1915 (1979).
39. S.W. Parr, *The Classification of Coal,* Bulletin no. 180, Engineering Experiment Station, University of Illinois, Urbana, 1928; *Annual Book of ASTM Standards,* vol. 05.05, Method D 388.
40. J.G. King, M.B. Maries, and H.E. Crossley, *J. Soc. Chem. Ind.*, *55*, 277T–281T (1936).
41. B.C. Parks, "Origin, Petrography, and Classification of Coal," in H.H. Lowry, ed., *Chemistry of Coal Utilization,* supplementary vol., Wiley, New York 1963, pp. 29–34.
42. *Instruction Manual, Fisher Model 490 Coal Analyzer,* Fisher Scientific, Pittsburgh, 1980, pp. 25–38.
43. E. Eickenberg, *J. Coal Qual.*, *2*(2), 35–36 (1983).
44. *Instruction Manual, MAC-400 Proximate Analyzer,* LECO, St. Joseph, MI, 1983, pp. 15–23.
45. Draft of "Proposed Test Methods for the Instrumental Determination of Carbon, Hydrogen, and Nitrogen in Laboratory Samples of Coal and Coke," American Society for Testing and Materials Committee D-5, Philadelphia, 1985.
46. K.A. Mellenger and W.T. Welch, *J. Coal Qual.*, *1*(2), 32–35 (1982).
47. C.M. Earnest, and R.L. Fayans, "Recent Advances in Microcomputer Controlled Thermogravimetry of Coal and Coal Products," *Perkin-Elmer Thermal Application Study 32,* Perkin-Elmer, Norwalk, CT, 1981.
48. M. Ottaway, *Fuel, 61*(8), 713–716 (1982).
49. *Instruction Manual, AF-500 Ash Fusion Analyzer,* LECO, St. Joseph, MI, 1983, pp. 15–18.

INDEX